TOWN AND HINTERLAND
IN DEVELOPING COUNTRIES

Town and Hinterland in Developing Countries
Perspectives on Rural-Urban Interaction and Regional Development

Milan Titus and Jan Hinderink (eds.)

HN
981
.C6
T68x
1998
west

THELA · THESIS

ISBN: 90.5538.024.5
NUGI: 671

© Milan Titus/Jan Hinderink (eds.), 1998

Cover design: Winneke Hazewinkel
Lay-out: Paula Duivenvoorde and Anneke van der Loo

All rights reserved. Save exceptions stated by the law no part of this publication may be reproduced, stored in a retrieval system of any nature, or transmitted in any form or by means, electronic, mechanical, photocopying, recording or otherwise, included a complete or partial transcription, without the prior written permission of the publisher, application for which should be addressed to the publisher: THELA THESIS, Prinseneiland 305, 1013 LP Amsterdam, the Netherlands. Phone: +31.20.6255429; Fax +31.20.620.3395; E-mail: Office@thelathesis.nl

ISBN: 90.5538.024.5

Table of Contents

Preface		
Introduction	Jan Hinderink	1
Chapter One	*Paradigms of Regional Development and the Role of Small Centres* Jan Hinderink & Milan Titus	5
Chapter Two	*Third World Urban Production and Employment Structures* Arie Romein & Milan Titus	19
Chapter Three	*Town and Hinterland in Central Mali: The Case of Mopti* Ali de Jong & Annelet Broekhuis	35
Chapter Four	*Small Urban Centres in Central Mali. A Study of the Role of Service Centres in Rural Development* Pieter van Teeffelen	65
Chapter Five	*The Absorption of Migrants and Their Living Conditions in Two Swaziland Towns* Cor van der Post	91
Chapter Six	*Small Towns, Labour Markets and Migrant Absorption: an Example from Northern Costa Rica* Arie Romein	111
Chapter Seven	*Reception Centre and Point of Departure: Migration To and From Nuevo Casas Grandes, Chihuahua, Mexico* Tine Béneker & Otto Verkoren	129
Chapter Eight	*Town and Hinterland in Central Java, Indonesia* Milan Titus & Alet van der Wouden	143
Chapter Nine	*In the Shadow of Yogyakarta? Rural Service Centres and Rural Development in Bantul District* Henk Huisman & Wim Stoffers	179
Chapter Ten	*The Small Town Reconsidered* Milan Titus	203
Bibliography		233
Appendices		251
List of Figures		259
List of Tables		261
List of Authors		263

Preface

This book presents the final results of a long-term and multilocational research programme on the structure and role of small towns in developing countries, carried out by the Department of Human Geography of Developing Countries in the Faculty of Geographical Sciences of Utrecht University. This programme has produced a large number of academic publications, reports and dissertation studies since the second half of the 1980s, but so far lacked a comprehensive discussion of its methodological and theoretical underpinnings, as well as a comparative discussion of its main research findings. Although it is impossible to present a complete picture of the various case studies of the programme, an attempt is made to offer at least a representative selection of its most relevant topics and case studies. Like any compilation study this book is the outcome of the selection of a limited number of cases, sometimes involving difficult choices and considerable textual condensations. The editors, therefore, are grateful to the contributors of this book who were often compelled to show themselves flexible and lenient in face of the many editorial requirements.
Last but not least our acknowledgements are extended to the technical staff supporting the editing and production of this book, i.e. Mrs. Paula Duivenvoorde and Mrs. Anneke van der Loo. Both have muddled through piles of incongruous texts and tables, and have been plagued by sheer unsolvable lay-out problems without losing their temper.
Finally, we should mention the staff of the Cartographic Laboratory who have (re)drawn many maps and diagrams in this book, as well as Mrs. Gina Rozario for her valuable contributions in correcting and adjusting several texts.

Jan Hinderink en Milan Titus

Introduction

Jan Hinderink

In 1983, a research team in the Faculty of Geography of Utrecht University started a programme of comparative studies on regional development in a number of countries in Sub Sahara Africa, Latin America and South-East Asia. Various considerations influenced the selection of the regions of study, such as already existing contacts with local universities and research institutes, cooperation with regional development organizations, and research experience obtained in the area by members of the team. All selected regions shared an important characteristic, viz. their increasing functional integration in the (inter)national political and space economies. Multifarious factors were thought to underly such integration. In the studies undertaken in the framework of the research programme special attention was paid to three important integrating factors: the commercialization of agriculture, the rural-urban interactions, and the sectoral and regional policies pursued by national governments and regional organizations. The aim was to gain insight into the nature of this increasing functional integration and its consequences for the production and living conditions of the populations concerned.

An important subject of research for the participants in the programme was the hitherho neglected role of small towns in the regional development process. This implied a profound understanding of their production and employment structures and their external linkages. At the time when the programme started relatively little attention was being paid to such lower order centres in the urban hierarchy, contrary to primary cities and large regional urban centres. The few studies that focused on small towns and the relations with their hinterland were found to have quite different and sometimes even contradictory views of their potential development role. This divergence in appreciation of small towns appeared to be partly caused by differences in theoretical approach and development perspective. Another reason was the insufficient attention paid in most studies to the characteristics of the regional context which may affect the functioning of urban centres, including small towns. The principal reason, however, was the poverty of empirical data that prevented a better understanding of the role of small towns.

The Small Town Research Project

Considering the importance of the subject, the existing controversies in the literature and the lack of empirical data, the research team decided to design a special research project on the structure and role of small towns in the selected regions of study. These regions differed in many respects: ecologically, demographically, economically, and socio-politically. Together with the various types of sectoral and regional policies pursued by governments and (semi)public bodies, this differentiation in regional context was supposed to be an important factor for possible differences found in the functioning of the small towns under investigation. Special attention was paid to their economic base and production structure, the dynamics and absorptive capacity of their labour markets for locals and migrants alike, their relations with the hinterland, and their interaction with higher order urban centres. Between them the cases were sufficiently differentiated to allow comparison at a higher level of generalization. In this way the research team hoped to arrive at a better understanding of the actual functioning of small towns and of the conditions under which these may play a positive role in the development of their hinterland.

The methods of data gathering in all case studies of small towns were basically the same. The collection of all types of secondary data relevant to the research took place at the national, regional and local levels - from ministries, (semi) public and private organizations, to libraries. Primary data were collected in the field by means of (sample) surveys among heads of households and owners/managers of enterprises, and by structured and open interviews with target groups and key informants. This primary data collection was carried out by graduate students from Utrecht University, and sometimes by students or staff from local cooperating institutions, under the strict supervision and in close cooperation with the members of the research team. However, no uniform 'blueprint' research methodology was possible because of the great variety in fieldwork conditions between and within the selected regions of study, such as e.g. with respect to accessibility, means of transport, diversity in cultural norms and social organization, bureaucratic rigidity and political control.

Yet, there was a common approach to the subject of research in all studies. Irrespective of differences in regional context, the attention was focused on the political-economic dimension of production and employment and the type of government policy pursued, and in the analysis conceptual models were used which had been developed by the members of the team in the course of their research. Such a common approach was thought to be essential for a proper understanding of the functioning of lower order urban centres in relation to the (changes in) production and living conditions of the population in small towns and their hinterland and for drawing valid general conclusions on the role of small towns in the regional development process.

The Contributions to this Book

The themes indicated above are elaborated in the various chapters of this book. The opening chapter by Hinderink & Titus deals with paradigms of regional development and the role of small centers. It reflects the differences in theoretical approach and development perspective as found in the literature. It also presents the view of the research team on interdependent development and the dialectics of the functioning of small towns. This view takes account of the divergent regional and political economic contexts of small town development and focuses particularly on its role in densely populated and more commercialised rural areas.

The conceptual models which are basic to this view are discussed in the following chapter by Romein & Titus. These models refer to urban production and employment structures of small towns and have been further elaborated in order to facilitate the analysis of the dynamics and absorption of small town-labour markets. The theoretical and analytical approaches adopted in this chapter reject the former dichotomous models and start from a much more flexible framework based on both structural and form characteristics of modes of production.

The next eight chapters present case studies of small towns in different regional contexts in Sub-Sahara Africa, Latin America and South-East Asia, The first two case studies are both situated in semi-arid and sparsely populated regions, viz. Central and South Mali. The case of Mopti as discussed by De Jong & Broekhuis, shows the impact of ecological constraints on incipient agricultural commercialization and economic growth, as well as their effects on urban production and living conditions and rural-urban relations of a Sahel town. In spite of considerable government efforts in improving the economic base and servicing role of the town, the relations with its rural hinterland did not improve significantly. Better ecological and economic conditions prevail in the more commercialised and functionally integrated southern part of Mali. However, as Van Teeffelen shows in this fourth chapter here too, the functioning of small towns as service centres and the use made of public services are still severely conditioned by such regional characteristics as poor physical and social infrastructures, low population density and purchasing power, and a vulnerable economic base.

The pervasive influence of the Republic of South Africa predominates the regional setting of Swaziland with its dualistic economy and high degree of economic dependence on its big neighbour. This political-economic situation is of major importance in understanding the urban employment structures and the absorption of labour migrants in the small towns investigated in the fifth chapter by Van der Post. These topics are also the main subject of study in chapter six by Romein, dealing with a small town in humid tropical and scarcely populated Huetar Norte, an agricultural colonization region in the north of Costa Rica. But the outcome here is quite different from the Swaziland case, showing both signs of migrant absorption and integration, and increasing bypassing of the town of Ciudad Quesada. This becomes comprehensible if account is taken of the completely different political-economic situation of Costa Rica as compared to Swaziland - both at the regional and the national level.

The case study in chapter eight by Beneker & Verkoren attempts to make clear the regional development and employment role of lower order urban centres in the semi-arid Northern Mexican border zone. The study deals with the interaction of a lower order centre (Nuevo Casas Grandes) with the regional urban system, focusing on its dynamic role as an employment centre in the context of rapid industrialization due to the establishment of foreign (United States) capital.

The last two case studies analyse town-hinterland relations in two densely populated and intensively cultivated and commercialised agricultural areas of Central Java, Indonesia. These are the more peripheral Serayu Valley and the district of Bantul, close to Yogyakarta city. The Serayu case study by Titus & Van der Wouden focuses again on the production and employment structures of three small towns, as well as on their functional relations with the rural hinterland areas and with higher order centres. Special attention is paid to the role of government interventions and development policies in shaping these rural-urban and inter-urban relationships. The Bantul study by Huisman & Stoffers analyses the small towns' role in rural development as service centres and their interaction with the regional metropolis Yogyakarta. Here, attention focuses on the issue of critical thresholds for establishing and maintaining public services in small towns in the face of changing rural production and living conditions.

In the final chapter conclusions are drawn with respect to production and employment structures in various types of small towns and to the nature of their hinterland relations in different regional and political-economic settings. On the basis of the findings of the various case studies an attempt is made to critically examine the interdependent view on regional development and the role of small centres that was developed in the course of this research project. Finally, an attempt is made to examine the main policy implications of the research findings.

CHAPTER ONE
Paradigms of Regional Development and the Role of Small Centres[1]

Jan Hinderink & Milan Titus

In discussions on regional development 'small centres' - i.e. the lower order centres in the urban hierarchy - and their role in the development process constitute an important subject of debate. Opinions differ as to their possible contribution to the spread of modernization, their impact on the development of the rural areas, their functioning as service centres, and their role in damming the rural exodus. These differing opinions are influenced by divergent paradigms of regional development - whether explicitly stated or not. This calls for a critical analysis of the relevance of existing paradigms and a search for other modes of interpretation which might increase our understanding of problems of development at lower geographical scales and of the functioning of small centres in particular. Such understanding may also help us to rectify possible misconceptions about the importance of small centres in regional development planning.

Existing Paradigms of Regional Development

The varying opinions on the role of small centres are, as we indicated, influenced by divergent paradigms of regional development which underlie the various arguments set forth. Three distinct paradigms can be distinguished (cf. Hinderink, 1983). The first considers development of regions as a function of national (economic) development. Great value is attached to regional planning as an instrument in the development process. This paradigm, which might be labelled the 'functional regional development' one, has up till now in a large measure influenced regional policies in many Third World countries.

A second paradigm, in many respects contrary to the first, might be described as the 'dependent regional development' one. It is a critical reaction to the optimistic assumptions of the 'functional' interpretation of regional development. Whereas development is again seen in terms of national (economic) aims, (regional) development, according to this paradigm, can only be brought about through structural change which transforms the existing relations of production and dependency. Therefore, the policies pursued should not only be geared to economic growth but should affect the existing political-economic order at the national, regional and local levels as well. A later version of this paradigm, which also emphasizes political-economic considerations, might be called the 'territorial development' one. This paradigm conceives of regions as territorial entities which should in the first place develop in their own way and according to their own interests. Consequently, regional development is seen a a process 'from below', which brings about an improvement in living conditions within the territory concerned.

In the following sections we will discuss the theoretical foundations and the main tenets of the various paradigms set out above. Special attention will be paid to the interpretations of the role and functioning of small centres and the different importance attached to these.

In the 1960s we find policy-makers and planners becoming more interested in the spatial dimension of development and the regional aspects of economic policy and planning. This attention reflects a growing realization of the short-comings of the policies pursued till then and characterized by a narrow concern for economic growth and, consequently for the allocation of public funds to the assumed 'dynamic' sectors of the economy, as well as by the concentration of investments and accumulation of capital in the metropolitan areas. This realization brings politicians and planners to reconsider priorities in formulated policy. Public funds are no longer merely allocated on the basis of national criteria but on regional ones as well. Regional planning is increasingly seen as an important instrument to realize regional development objectives. Attention focuses not only on the solution of problems of the rapidly growing metropolitan areas and commercialized agricultural regions, but also on those of economically stagnant zones, peripheral areas and resource frontiers. However, the development objectives of regional policy are subordinated to national (economic) aims. Regions are considered as open systems and their development must promote their functional integration into the national economy (Hilhorst 1971; Friedmann & Weaver 1979).

What is the theoretical foundation of this view of regional development, labelled as 'functional'? Important sources of inspiration are the theories of economic polarization, location, and central places, as well as the various modernization theories. It is not possible in this context to enter into detail and discuss the individual characteristics of these theories. Suffice is to say that all tend to underrate the political dimension of development, all attribute a positive role to government action and public organizations in the development process, and all show a distinct policy bias in favour of cities and urban areas.

Regional economic policy and planning, influenced by the paradigm of 'functional regional development', find expression in two major strategies, viz. growth centre and rural service centre strategies, which are defined respectively in terms of (industrial) production growth and in terms of distributive trade, administration and services (Mosher 1969; Dusselkorp 1971; Doherty 1975; Appalrayu & Safier 1976; ESCAP 1979, 1990). These strategies are assumed to lead to an effective use of regional resources and an increase of productivity and output, to social change and economic transformation, and to a decrease of regional inequalities in development. Actually, the two strategies are found to be partly overlapping, both in the literature and in the implementation of planning. This is not surprising, considering the fact that a developed industrial growth centre may generally perform the functions of a central place as well.

A number of countries in Latin America, Africa and Asia purposefully attempted to implement growth centre strategies in regional development. Their objective is to stimulate production activities, often in the framework of import-substituting industrialization, which trigger off economic growth and (re)structure the regional economy. The emphasis is mostly on large-scale capital investments and their concentration in urban areas in view of supposed economies of scale and location. Centres qualifying especially for such investments are those which function as inter-regional foci of production and trade, usually cities with one-hundred thousand inhabitants and more. It is assumed that the spread effects of polarized growth in these medium-sized centres will have a stimulating effect on the smaller ones within the region. Economic linkages with the regional centre will gradually transform the smaller ones into lower order growth foci and make them 'instruments of modernization' for their immediate surroundings (Misra & Sundaram 1978). The development of medium-sized centres will also counter primate city dominance and bring about an 'optimal urban hierarchy'. Such a hierarchy is thought to facilitate the diffusion of innovations to the lower order smaller centres, thus enabling these to perform in an adequate way their intended role in the development and modernization of their rural hinterland areas (Johnson 1970; Berry 1972; Misra & Sundaram 1978; Rondinelli & Ruddle 1978; Rondinelli & Evans 1983).

However, in developing countries the expected spread effects often appear to be of little consequence for the smaller centres. A major reason is that economic linkages are especially strong with extra-regional and overseas clients and suppliers. In this respect, Mabogunje has already pointed out the limited results of growth centre strategies based on import-substituting industrialization and the requirements which condition their possible success in Third World countries (Mabogunje 1978). If the regional policies pursued also aim at the development of smaller centres for the benefit of the rural areas, it seems more appropriate to invest directly in such countries. The question is whether to invest in all of these or only in centres of a specific order, or in those enjoying a particularly favourable location. Especially in regions where motorized traffic is already rather well developed it seems less desirable to promote the emergence of a dense network of service centres. In less developed regions, however, the stimulation of 'hintertowns' or a type of 'intermediate settlement' is advocated as a supposedly necessary link between the rural areas and the higher order urban centres (Manshard 1977). In a number of developing countries attempts have been made to stimulate in a planned way the development and distribution of such rural centres. Their role and importance is no longer seen as derived from the spread effects of polarized growth in the medium-sized regional cities. On the contrary, these rural centres are considered 'engines of growth' in their own right and for the benefit of their rural hinterland. Their development as market and service centres helps to increase the productive capacity of the rural producers and promotes the commercialization and specialization of agriculture in the framework of national economic growth (Mosher 1969; Dusseldorp 1971; ESCAP 1979). The planning and proper functioning of these rual centres is thought to contribute to an 'effective' integration of the rural population in the national economic and political order. Moreover, it is assumed that the development of small centres increases local employment and, in this way, helps to stem the rural exodus to the metropolitan areas (Johnson 1970; Southall 1979). As illustrations of policies which are directed at small centre development in selected regions we may point to the 'Agrovilles' in Pakistan, the 'New Towns' in Eastern Malaysia, and the 'rural centres' in the 'intensive rural development zones' of Zambia. In countries like Indonesia, Kenya and Colombia, the central place model has influenced regional planning policy in the spatial structuring of the national territory (Taylor 1974; Soegiyoko & Sugiyanto 1976; Appalrayu & Safier 1976; Taylor & Obudho 1979).

From the end of the 1960s onwards the 'functional regional development' paradigm comes to be increasingly criticized. The criticism is levelled both at the naive optimistic assumptions underlying the industrial-urban biased growth centre strategy and at the instrumental character of regional planning itself, implemented as spatial engineering. The theoretical foundation and source of inspiration is dependency thinking in its various forms. Critical analyses are actuated by the disappointing results of many regional development projects, particularly by the absence of any substantial spread of economic growth. These analyses make clear that the expected social and economic transformations have not come off and reveal that in many cases the existing power structures have even been reinforced by the strategies pursued. Growth centres and central places are increasingly seen as foci which are encapsulated in the existing political-economic order and reinforce it. Their growth and functioning is assumed to serve vested interest.

This 'dependency' view of regional development does not deny the potentially positive role of small centres as growth foci and central places of a lower order. However, with the prevailing national power structure of most Third World countries and in the present international political and economic framework they are seen as 'vanguards of exploitation'. These vanguards function as the tentacles of the large urban centres, which still supposedly perform their former colonial role of drainage mechanism and intermediate for the benefit of the western industrial powers. In various studies on the role of small centres in Kenya, Sudan and Zaire their weak productive functions are emphasized. Their distributive functions turn out to serve particularly the interests of a small political and commercial elite at the national level and associated international interest groups (Kabwegyere 1979; Ahmed & Rahman 1979; Schatzberg 1979). As far as Zaire is concerned it is said that the fewer of these small centres there are, the better it is for the mass of the population in view of their negative impact on rural development (Schatzberg 1979). A fundamental solution to overcome these problems of drainage and exploitation is seen in a radical change of the existing political and economic structures in developing countries and in policies geared to self-reliance and to the needs of the mass of the population.

Partly linking up with the 'dependent regional development' paradigm, the 'territorial development' one also pays attention to the existing divergence in interests and needs, in short to the political dimension of such development. On the other hand, this paradigm distinguishes itself from both the 'functional' and 'dependent regional development' ones by its strong emphasis on the negative effects of centralizing tendencies and bureaucratic control to the detriment of local action and participation at the grassroot level. The theoretical foundation of this paradigm is rather weak. This appears among other things, from the vagueness of such a key concept as 'selective territorial closure' which is seldom concretely defined in terms of scale. The sources of inspiration are the dissatisfaction at a narrow materialistic concern and top-down approach to development, and a strong sympathy for the ideals of self-reliance and 'small is beautiful'. The 'territorial development' paradigm rejects the conception of the region a an open system and the subordination of regional development objectives to national (economic) ones. Its supporters advocate the use of the region's resources for the benefit of the regional population, the development of an integrated and diversified agro-industrial economy geared to regional needs, the decentralization of planning and decision-making, as well as the grassroots participation and co-operative action at the local level (Lo & Salih 1981; Stöhr 1981; Weaver 1981).

The most concrete example of regional development from a territorial point of view is the model of agropolitan districts' (Friedmann & Douglass 1978). This model relates to the development of territorial entities comprising fifty to one-hundred and fifty thousand people and including a service-cum-industrial centre. Capital accumulation and re-investments should take place as much as possible within these agropolitan districts, and selective territorial closure should prevent the leakage of generated wealth. Priority should be given to small-scale and labour-intensive rural productive activities, and export-based activities 'should be promoted only to the extent that they lead to a broad increase in living standards of the population of the territorial unit' (Stöhr 1981, p. 66). Impulses to development are assumed to start from below and to 'filter up' from local to regional and, finally to national levels. Accordingly, policy and planning should be adapted to agropolitan reality.

It remains to be seen whether selective territorial closure and decentralization on the basis of a local and broadly-based participation are feasible propositions, given the political, economic and administrative reality in developing countries. Up till now, the territorial development paradigm has yielded hardly any tangible results in the form of detailed proposals and worked out plans. However, its attention to administrative structures and types of decision-making is important, especially with regard to the role and functioning of small centres. From a territorial point of view centres are not only conditioned by a country's political and economic order, but also by the existing hierarchical patterns of centralized control and by bureaucratic structures. However, such patterns and structures are not specific to any particular political and economic order.

Critical Reflections on Existing Regional Development Paradigms

A crucial question with respect to the paradigms discussed above is their - tacitly assumed - universal validity. This question is posed, considering the inductive nature of conceptions about regional development as influenced by these paradigms. This often leads to an incomplete analysis of the multifarious and dialectic character of development processes under (peripheral) capitalistic conditions. A certain mechanistic rigidity seems to underlie such analysis, even if political-economic considerations are taken into account.

Sure enough, analyses based on the 'dependent regional' and 'territorial development' paradigms primarily seem to emphasize the exploitative and inequality-increasing character of the capitalist expansion process. This explains that hardly any attention is paid to the dialectics of internally and externally induced processes which in the long term (may) have contradictory effects.[2] The prevailing ideas of the dependent regional and territorial development paradigms are that only radical transformation of the existing economic and political power-structures or selective territorial closure and decentralization of control and decision-making may stop ongoing exploitation and deteriorating terms of trade in peripherally integrated regions. Moreover, both paradigms hold that regional underdevelopment is the necessary outcome of the capitalist expansion process. This is characterized by the international and internal division of labour, based on types of accumulation and circulation which are conditioned by the international and national metropoles (cf. Weaver 1981, pp. 35-87). Through unequal exchange, control of technology and political dominance surplus is transferred to the 'metropoles', increasing the development gap with the periphery. Surplus transfer takes place between social classes and between spatial configurations, the latter usually being defined in terms of regional and urban hierarchies (Friedmann 1966; Stöhr 1975; Morris 1981). Both types of surplus transfer overlap and reinforce each other because of the concentration of a bureaucratic-capitalist elite and middle class in the urban centres - the lower order ones constituting the ultimate chain in a hierarchical system of exploitation. The consequences of the penetration of capitalist (corporate) forms of production in the rural areas make them increasingly dependent on large cities and foreign countries as market outlets and supply centres for inputs and consumer goods (Lo & Salih 1978).

Small-scale trade and rural industries decline under the impact of growing external competition. Capitalist penetration promotes cash-cropping, often to the detriment of the cultivation of food crops, and induces wealthier farmers to adopt capital-intensive techniques and new crop varieties. Consequently, pressure on land increases, while the number of landless augments. The final result is 'agrarianization' of the rural economy and growing out-migration of pauperized peasants and landless to the larger urban centres which - metaphorically spoken - are offered the bill for their privileged position (Lee 1981).

Interpreted in this way, capitalist expansion and integration is seen as a unilinear and causative process of progressive impoverization of peripheral areas. It would appear that this mechanistic interpretation is the regional variant of the original 'immiserization' theory, based on class exploitation. However, this interpretation does not take into account one of the most fundamental principles of Marxist analysis, viz. the dialectic character of capitalist development itself. As a matter of fact, in the Third World too, the dynamics of capitalist expansion and accumulation lead to a continuous exploitation of new resources and the opening-up of new markets (Roberts 1978). This is manifested in the progressive integration of peripheral regions in the national and international economies. It finds expression in a growing interdependence of regional economies, reinforcing hierarchical relations as the 'Centre' retains its dominant position. Yet, this does not prevent or hamper an increase in productivity and average income or, it has been observed, an overall growth of the regional product (Gilbert & Gugler 1982). Through the introduction of new crops, organizational and technological innovations and the exploitation of new resources the increase in total production can be such that even with a relative increase of surplus transfer there may still be a substantial absolute growth of the available surplus within the region.[3] This is particularly true of the growing class of wealthier farmers and traders who do not only spend their increased incomes in a consumptive way, but also invest in so-called non-farm activities (Anderson & Leiserson 1980; Freeman & Norcliffe 1984; Rietveld 1986; Herlaar & Sonnema 1987). These activities concern the processing, marketing and transportation of the increased agricultural production surplus, the increased trade in consumer goods and production inputs, and the expanding repair services and credit supply to meet the growing regional needs. In this way a process of economic growth and diversification may start in regions where initially the consequences of the 'Green Revolution' and functional integration in the national economy were primarily characterized by increasing agricultural commercialization and specialization and the decline of traditional non-farm activities. Employment opportunities may grow through increasing intensification in agriculture and by an expanding job market outside the agricultural sector. This development is clearly demonstrated in the more commercialized areas of Africa and in large parts of South and South-East Asia, where growing population pressure goes hand in hand with increasing levels of agricultural productivity without any decline in real agricultural wages. In several highly commercialized areas labour shortages may even be observed during peak seasons in the agricultural calendar (Livingstone 1986; Collier 1982).

The conditions for this so-called polarization reversal in rural development processes constitute the central theme of the territorial development paradigm. As has been pointed out above, this paradigm holds that a transformation of negatively valued developments can only be brought about if policies are pursued which aim at more autarchy and territorial closure of the regional economy and which lead to far-reaching decentralization in political and administrative respects. It is quite remarkable, however, that recent research has shown that this 'polarization reversal' also occurs in a number of regions where such policies have not been implemented. Especially in the heavily populated, more commercialized and integrated rural areas of Kenya, the Punjab, Central Thailand, West Malaysia, Java and Central Luzon developments seem to materialize which are only assumed to take place under the conditions stipulated by the above defined paradigm (cf. Collier et al. 1982; Jayasuriya & Shand 1983; Freeman & Norcliffe 1984; Jones 1984; Van Oosterhout 1985).

Even more than the direct stimuli given by governments through special sector programmes and capital expenditures for institutional and infrastructural development, it is the indirect effect of the transformation process which takes place in the rural areas and the urban centres that creates additional and/or substitutional employment. It is true that such employment is primarily found in the informal or petty commodity sector. It is also true that this type of employment is more developed in the urban centres than in the countryside. This has resulted in an increasing circular labour mobility between the rural areas and the urban centres in Sub-Saharan Africa and in South and South-East Asia (Meilink & Van Binsbergen 1978; McGee 181; Hugo 1982). This form of labour mobility may therefore be considered as an adequate response to an urban-biased type of development. For it seems that in this way part of the drained surplus to the urban centres can filter back to the rural areas without the rural economy having to suffer the complete removal of the circular labour force involved. Besides, the more marginalized rural dwellers may increasingly benefit from infra-structural improvements and modern transportation technology in circulating between their home areas and the towns. In doing so they retain the security of their village and avoid the higher living costs in case they and their families settled permanently in town. Together with gradually growing employment opportunities in the non-farm sector, the option of circular labour migration makes it possible that rural proletarianization may to a large extent proceed without substantial dislocations and serious social effects 'in situ'.[4] Moreover, according to some authors, complete proletarianization is relatively limited because most marginalized operate as selfemployed or are absorbed as family workers in the informal or peasant sectors (Roberts 1978; Livingstone 1986).

This is not to deny that polarization reversal often implies very marginal forms of participation in semi-capitalist production structures. Generally, the activities concerned are of a distinct 'supply-push' character - i.e. activities which are performed at almost any price, out of the bare necessity to find a living. But such petty type activities should certainly not be regarded as superfluous in a peripheral capitalist production system. They often demonstrate a higher flexibility and persistence than what might be expected in view of their weak capital and technological basis (Roberts 1978; McGee 1981). Low costs of production through self-exploitation of labour make these activities 'functional' - both in meeting the needs of low income groups and in enabling the corporate sector to save on expensive overheads and costs of distribution. This explains why petty commodity activities are quite resistant to outside pressure. They also enable the corporate sector to keep real wages low. In this way, national firms are in a position to maintain themselves in the domestic market in spite of foreign competition. Occasionally they may even penetrate in foreign markets. Therefore, the internal dynamics of the capitalist sector are actually supported by the petty producers in the informal and peasant sectors. On the other hand, these petty producers depend for their survival to a large extent on the circulation of money and commodities which is again generated by the capitalist sector. This shows that the internal dynamics of peripheral capitalism and of its spatial expansion are considerably more complex than the various existing paradigms would lead us to believe.

The Dialectics of Development: Small Centres Revisited

We would now try to revise the conventional analysis of the role of small urban centres, taking into consideration the framework set out above. According to the functional and dependent regional development paradigms, this role can be of only one kind: either stimulating and innovating or parasitic and exploitative. These paradigms do not consider the possibility of a change in the functioning of such centres or, simultaneously, a partly stimulating and partly exploitative role in the context of a dialectic process as described before. Moreover, recent research makes clear that the role of small centres does not condition the development of their hinterlands - either in a positive or in a negative way - to the extent that is assumed by these paradigms. Research carried out by geographers in Sub-Sahara Africa and South-East Asia has shown that the most important impulses for the development and transformation of the rural periphery come directly from the higher order centres or are determined by central governments and world market mechanisms (Harts-Broekhuis & Tempelman 1983; Titus 1982; Titus et al. 1986; Hinderink & Sterkenburg 1987). The result is that local or regional service centres are often completely bypassed. This applies particularly to centrally guided government programmes in the field of food production, cooperatives, rural industrialization, credit supply, electrification and infrastructural development. In this case, the district and rural service centres are only allotted a subordinate and supporting role. It may also be true that socio-economic changes in the rural hinterland lead to the loss of functions and 'by-passing' of these centres. Especially when agricultural intensification and innovation bring forth increasing production surpluses and economies of scale, traders from higher order centres may find it attractive to short-circuit the relatively small-scale collective trading chains of the local service centres

(Logan & Missen 1979; Titus et al. 1986). Besides, growing purchasing power and changing consumer behaviour of the rural middle class may also lead to the bypassing of local service centres, particularly for durable consumer goods and specialized services. Under these conditions such centres may not be able to function as centres of innovation and modernization. It is also clear that under these conditions their exploitative role vis-à-vis their hinterland area will remain relatively limited.

An important conclusion that can be drawn on the basis of the ongoing research on the role of small centres is that the quality and diversity of their functions depend on the developments in their hinterland instead of the other way round (De Jong 1983; Harts-Broekhuis & Tempelman 1983; De Jong & Van Steenbergen 1984; De Jong & Ligthart 1986). When the rural economy is more commercialized and based on the surplus production of independent producers, the economy and functions of small centres tend to be more differentiated and more resistant to the dominance and bypass effects from the higher order centres.[5]

The small urban economy's reliance on its rural hinterland does not, of course, exclude a predominantly exploitative relation to the detriment of the hinterland. Especially at this low level, such an exploitative relation manifests itself in the unequal exchange between rural and urban-produced goods and services, as well as in the drainage of capital through a repressive tax system, unfavourable tenancy conditions and oppressive rates of interest on credits supplied. But for the rural hinterland the importance and the negative effect of this dependent relation with rural service centres is generally subordinate to the importance and effects of its relations with markets and higher order centres outside the region. The reason is that the surplus appropriated by the rural service centres is still consumed or spent within the region. Moreover, the direct dependence of these centres on externally induced developments in their hinterland limits the exploitative character of their relations with the rural area at the penalty of being bypassed. This explains why it is just this type of urban-rural relations which may be changed into a more symbiotic way through the growth of small urban functions which support the rural production and transformation process. For it would seem that sooner or later such a re-orientation of the relations with the rural hinterland may turn out to be more lucrative than the continuation of the existing exploitative ones. These, after all, are hampering the growth of the rural production surplus and are setting a limit to the surplus which is available for the service centres themselves.[6]

The small centres' function of meeting the growing needs and possibilities of their commercializing hinterland finds spatial expression in the expansion and diversification of daily and weekly markets and in increasing activities in the fields of retail and collective trade, transportation, repair and agricultural services and processing. Simultaneously, the local government, educational and health services may also improve and expand - although within the limits set by the dominant higher order centres and the metropolitan region.

Contrary to what the dependent regional and territorial development paradigms would lead us to believe, the essential factor for development is not so much the halting or even the reversal of the rural-urban and inter-regional flows of surplus transfer (cf. Lo & Salih 1981, pp. 123-134). Of more importance is the condition that both in the rural areas and rural service centres sufficient surplus increase remains to set into motion and perpetuate a process of rural tansformation. This also implies that in these rural areas and centres a middle class emerges that is willing and able to invest surplus income in local production activities. Admittedly, these investments will initially be primarily directed to agricultural production. But they will also increasingly be attracted to non-farm activities (Freeman & Norcliffe 1984; Weyland 1984; Rietveld 1986). The principal reason being that usually the profitability of new investments in agriculture declines after the initial phase of sharply increasing productivity. This declining profitability compares unfavourably with the average higher profit margins and values added in the non-farm sector. Even under semi-monopolistic conditions the volume of sales in this sector provides acceptable profit margins.[7] This makes clear why the growth of these non-farm activities cannot any longer be explained exclusively in terms of a 'supply-push' response to the ongoing proletarianization of the rural population.

The conditions for bringing about a surplus increase in peripheral regions may in large measure be created by policies pursued at the national level. It is, however, doubtful whether these policies must be the ones as advocated by the protagonists of the territorial development paradigm. For it is difficult to realize in most developing countries such recommended measures as far-reaching decentralization of political and economic power, selective closure of markets and attempts at regional autarchy in the field of basic needs (cf. Friedmann & Weaver 1979, p. 195; Stöhr & Taylor 1981, p. 47). These measures are not only difficult to implement in view of their assumed negative effect on vested interests and, consequently, in view of the political resistance to be expected. What is even more important: they are also counterproductive to the development of an integrated market economy at the national level. Therefore, it might be far better and feasible to pursue policies which have already proven their value in stimulating small producers, e.g. factor price control, bottom prices for food and agricultural exports, selective credit supply, land reform and institutional improvements of the regional production structure in the field of administration, education, extension, marketing and cooperatives.

Such policies are also better geared to the possibilities of an open market economy. For in one way or another, this type of economy remains the most important driving force behind regional integration and transformation processes in most Third World countries. This fact is also acknowledged in the dependent regional and territorial development paradigms. However, these paradigms deny the intrinsic value of such an economy because they merely emphasize, albeit in different ways, its direct exploitative effects. Neither does the functional regional development paradigm provide an appropriate framework to understand the complexities of the market economy. Policies geared to promoting the developmental role of regional service centres may be of no avail if, under certain conditions, these centres only play a minor and subordinate role in externally induced transformation processes. The creation of growth centres at this low hierarchical level would go against all the 'laws' of dependence and dominance relationships and would risk ending in an outright failure. Another constraint in the implementation of such development policies in the prevailing capital shortage at this level. This makes it impossible for governments to mobilize sufficient resources for investment in small centres and to develop their 'critical mass', i.e. the economic complexity which is required for a proper growth centre function (Lo & Salih 1981, p. 10).

A more realistic option under these conditions seems a policy which takes into consideration the dominance of capitalist production relations at the regional level. Such a policy should be based on the regional development potential. At the same time attempts should be made to minimize as much as possible any leakage of the regional surplus. But these attempts must not be confused with such unrealistic propositions as selective territorial closure or regional autarchy. Attempts should rather be geared to securing a sufficient surplus for producers in the region - e.g. through guaranteed prices, land reform, improved marketing and the creation of an attractive investment environment for locally accumulated capital of independent producers. The scope of this chapter does not allow us to go much deeper into this subject.

Finally, we would like to say something about the potential developmental role of small regional service centres. Rondinelli (1983) has pointed out that this role may be reinforced considerably through carefully designed policies and planned investments by national and regional authorities. Special attention should be given to productive functions in the fields of marketing, transport and processing of products from these centres' hinterland, and to the expansion of retail trade, credit supply and agricultural extension. In addition, administrative, educational and health facilities should be improved to meet the increasing demand from the commercializing rural hinterland and to keep the generated capital within the region. In this way, employment opportunities in these centres will increase as well. Development along these lines may also stimulate circular mobility, thus contributing to the creation of additional sources of income for people in the rural areas.

Such a selective supporting role of small regional centres in the framework of a rural development policy is quite different from the role as envisaged by an urban-biased and costly growth centre policy. On the other hand, it would appear that such a selective supporting role corresponds to the agropolitan model as propagated by the territorial development paradigm (cf. Friedmann & Douglass 1978; Lo & Salih 1981). However, there are some fundamental differences. The policies promoting a selective supporting development role of small centres will be implemented in an open economy, whereas the agropolitan model presupposes territorial closure. Moreover, a development process as set forth above, implying a mixture of top-down and bottom-up initiatives and impulses, seems more realistic under the prevailing conditions in most Third World countries than the somewhat Utopian philosophy of bottom-up development which underlies agropolitan development thinking.

Towards a New Paradigm of Regional Development?

The ideas about regional development and the role of small centres as outlined above emphasize bilateral dependence relations at various levels or geographical scales, relations which are characterized by dialectic complexity. This view of regional development might be labelled a paradigm of interdependent regional development.

Its principal policy implications may be summarized in the following way. Agricultural intensification and innovation should bring about an increasing agricultural production surplus as the basis for a growing economy in (peripheral) rural regions. Constraints in the organization of production and social production relations which negatively affect production increase and income distribution must be overcome through institutional reforms. Problems of leakage of the regional surplus should be countered by specific policies in the field of pricing, factor price controls and marketing, and be adjusted to the conditions obtaining in the concrete regional context under consideration. The creation of a favourable 'investment environment' for local capital in such a context is an essential condition for bringing about the necessary polarization reversal in the rural and regional economy. Beside the stimulation of investment opportunities in the rural farm and non-farm sectors, this implies a reinforcement of the functions of regional service centres in the common interest of both rural and urban producers. Such a reinforcement would especially mean the improvement of commercial and non-commercial services in the fields of collecting and distributive trade, credit supply, processing, rural extension, education and health. Of prime importance in this process is the expansion of employment opportunities within and outside the regional context, with due acknowledgement of the role of petty commodity production and circular labour mobility. These policy implications of the interdependent regional development paradigm make clear that there is no simple blueprint for development. Innovation and stimulation, prevention of leakage and of exploitation - all these objectives cannot be met in a single short time span. To achieve these ends, policies must be pursued which are necessarily complex and which should be flexible enough to take account of changing conditions in the ongoing dialectic development process.

The question remains, however, whether (and where) the paradigm of interdependent regional development has any empirical validity. Or, to put it more modestly: does the reality of the Third World provide enough evidence to assume the paradigm's relevance? It has been pointed out before that processes of rural transformation as seen from this dialectic point of view are particularly found in the more densely populated, strongly integrated and commercialized rural areas of Africa and Asia, where the production structure is primarily characterized by small-scale selfemployed producers. The paradigm seems less relevant for highly commercialized agricultural regions which are dominated by latifundia or agribusiness. In such type of regions the leakages are too large, the obstacles for the emergence of a class of small selfemployed producers are too big, and the possibilities for economic diversification too weak. Other important conditions for the paradigm's relevance are: the political will to increase the surplus of rural producers and regional centres, as well as a well-developed physical infrastructure and access to external resources such as urban employment. It is therefore doubtful whether our model is applicable to regions with a weak infrastructure and resource basis and with a predominant tribal mode of production. These regions lack the basic conditions for a surplus production oriented to export and trade. Besides, such production is hardly feasible, considering the high cost of investment and the social, economic and demographic problems involved. Under these conditions an agropolitan model, based on intra-regional trade and regional self-consumption, may probably be more appropriate or at least less harmful as a development model. Seen from this perspective, it is quite surprising to find that the protagonists of this model consider it precisely most relevant for the densely populated and commercialized rural areas in Asia (cf. Friedmann & Douglas 1978: Lo & Salih 1981).

Conclusion

The theoretical viewpoints set forth above focus on important issues and contain many valuable elements. Because of their different orientation and emphasis they sometimes seem to conflict with each other and to lead to quite opposite conclusions about the development role of small centres. Another reason why conclusions may seem to contradict is that the argumentation is often based on a particular case and is nevertheless formulated in general terms. However, generalizations do not always convince and they may even confuse rather than improve our understanding of the role of small centres in regional development. What we need is empirical evidence. True, the evidence shows that, with improved transport and communications and with growing commercialization and specialization of production, regions become increasingly functionally integrated in wider political and economic networks i.e. both national and international. This is attended by a hierarchical structuring of space, resulting among other things, in the emergence and growth of small centres. But the actual role of these centres is difficult to gauge without an analysis of their concrete regional context, taking account of political-economic reality and the structure of control and decision-making. Such an analysis may also reveal that in a continent like Africa many regions are so poorly endowed and so little developed that there is hardly any marketable surplus. Whereas such regions may be well-integrated in a political and administrative respect, their weak economic base negatively affects the functioning of their centres. To put it otherwise, at a very low level of development the service function of small centres may be quite restricted and the nature of the services provided may be of no great use to the majority of the rural population. On the other hand, strong evidence has been found in the more advanced rural regions of Asia and Latin America of a more complex role of small urban centres in regional development. Therefore, generalizations about the role of small centres, based on whatever theoretical point of view, are hard to make. Due attention has to be paid to the level of their hinterland's development and to the prevailing political and economic conditions. The lesson to be learned then is that if we want to acquire insight into the potential developmental role of small urban centres, the regional context should play an important part in our analysis.

Notes

1. This chapter has been published earlier as an article in *Development in Change*, 1988, vol. 19, pp. 401-423.
2. It is true that both Friedmann & Weaver (1979) and Stöhr & Taylor (1981) have recognized dialectical principles in the regional development process, such as the polarity in functional and territorial integration, and the inherent contradictions between development from above and from below. However, this dialectic has little to do with the one inherent in the capitalist development process itself. In this process surplus transfer, accumulation, class formation, competition and the urge to innovate lead to a continuous process of increase of productivity, market expansion, capital formation and capital destruction (cf. Schumpeter 1942, on the process of creative destruction).
3. According to Gilbert & Gugler, an increase in relative exploitation together with a simultaneous absolute growth of the regional surplus characterizes the phase of monopolistic capitalism (Gilbert & Gugler 1982, p. 37). This phase has also begun in the strongly integrated and commercialized rural hinterland of Third World metropoles through the penetration of the activities of banks, commercial ventures and agribusiness. However, this penetration does not automatically imply the development of large-size agricultural enterprises.
4. Seen from this perspective it becomes clear why agrarian wages in such divergent regions as Java, the Punjab Thailand, Malaysia and Kenya do not show any real decline, despite an outflow of labour from a modernizing agricultural sector or increasing rural population pressure (White & Makali 1979; Jayasurya & Shand 1983; Livingstone 1986). Moreover, there is strong evidence that the growth of circular mobility and the ongoing

absorption of labour in the rural economy are explanatory factors for the slow increase of the level of urbanization in many countries of South and South-East Asia during the 1970s (cf. McGee 1981; Jones 1984).

5. Gilbert & Gugler also observe that small-scale export production by peasants is more inducive to the local growth of urban centres than the development of large-scale estates or mining activities (Gilbert & Gugler 1982, p. 42). This is because of the stimulating effect of a greater surplus for small-scale producers on the demand for consumer goods, agricultural inputs and services. This demand is met by local markets and services rather than by organized corporate supply and leads to a more balanced urban network.

6. This is not to say that this re-orientation in rural-urban relations will take place spontaneously and smoothly. Vested interests will be affected and may cause reactions from the traditional (rentier-capitalistic) elite and middle class in the service centres. The dominant position of this elite and middle class, however, may sooner or later be threatened by an emerging urban entrepreneurial class, whose interests coincide with those of a development-oriented government and an emerging rural middleclass (cf. Lefeber 1978, p. 8).

7. Higher profitability of non-farm activities seems to conflict with interpretations of such activities in terms of 'last resort' or 'refuge-type' ones (cf. White 1976). Actually, the non-farm sector is a very heterogeneous one. Although based on agricultural production, this sector takes up a strategic position between the peasant sector and the urban economy, often allowing for price-setting under semi-monopolistic conditions. This explains why non-farm activities may be an interesting field of investment for wealthier peasants. Even the marginal activities in this sector, generally less lucrative than agricultural activities if valued in terms of unit time, may still constitute a realistic alternative. The reason is that, if valued on a yearly basis and if account is taken of long working days, such activities ultimately yield a higher income than the season-bound agricultural activities.

CHAPTER TWO
Third World Urban Production and Employment Structures and the Small Town

Arie Romein & Milan Titus

Referring to the employment situation in Third World urban centres, Lewin wrote that

> 'the expulsion and migration of rural workers and the accelerated rate of natural population increase have led to the growth of the urban labour force seeking employment. Capital-intensive industry and the large-scale formal public and private tertiary sectors are incapable of absorbing more than a small share of the job seekers. Consequently, an increasing share of this growing labour force has to earn a living in activities defined as the informal sector, petty production or casual labour' (Lewin 1985, p. 129).

In other words, this increasing part of the urban labour force that remains untapped by modern, large-scale enterprises 'lives a scratching life outside the organised labour market' (Breman 1976, p. 10). Considering the case of the smaller towns, there is no reason to consider them as an exception to this general rule, for a large modern industrial sector usually has never developed in or around these towns while their labour force has grown rapidly over the past few decades. So far, only some research into the theme of economic and employment structures in small towns has been carried out (Titus et al, 1986; Davila & Satterthwaite 1987). Considering the growing interest in the role of small towns as regional growth centres and as alternatives to metropolitan development further research into their structure and functioning seems highly desirable.

Such new research implies that choices have to be made, with regard to both the theoretical premises and the empirical instruments of analysis. This chapter intends to give a brief overview of not only the main theoretical points of view, developed so far with regard to Third World urban production and employment structures, but also of the operationalisation of these viewpoints into the practice of empirical research. We hope that this overview may be useful as a guide-line to coming research into the structure and role of small urban centres.

Two Opposite Approaches

The structure of Third World urban economies has been under discussion since the beginning of the 1970s in both academic and policy-describing circles. A generally accepted postulate in this debate is the separation of urban economies into distinct segments. Each of the segments includes certain types of enterprises and working conditions of a certain quality range. Despite the general acceptance of this postulate, there is no consensus on the precise nature and characteristics of these distinct types of enterprises or working conditions. It was as late as 1991 that Uribe-Echevarría (1991, p.3) could conclude with regard to the most frequently studied segment, the informal sector, that 'few concepts are more vague, imprecise and misleading than this one'.

Two opposite approaches can be distinguished in the discussion on urban production and employment structures in developing countries: the dualist approach versus the structuralist approach. The debate on these two approaches takes place on both the empirical level (concerning the precise number of different segments and their respective characteristics and definitions) and the political-ideological level. The main themes of discussion on the latter level concern the nature and functions of the interrelations between segments and the part played by the national state in the development of urban production and employment.

A brief résumé of the debate on the two opposite approaches is presented in this section. These two approaches are considered ideal-types; the résumé does not enter into their internal variety in points of view, methodology or conclusions.

The notion that Third World urban economies are characterised by a segmented structure arose in the 1960s. In a study on the Indonesian town of Modjokuto published in 1963, Geertz distinguished between a firm-centred and a bazaar-centred economy. Eight years later, in 1971, Hart presented a paper, based on his research into the 'reserve army of urban underemployed' in Accra, Ghana. In his paper, he made a distinction between a formal and an informal sector. It was this dichotomy that became adopted by the ILO (International Labour Organisation) in 1974 in a study on urban (un)employment in Kenya. This report meant the start of further elaboration of the formal-informal dichotomy by the ILO into a true school of thought: the dualist approach or, as it is referred to by some authors (Moser 1978, p. 1042, Breman 1985, p. 52), the ILO approach.

Empirical research in different cities in the developing world has revealed considerable differences between the formal and the informal sector. Generally speaking, the enterprises within the formal sector are larger, are more capital-intensive and use more advanced technical equipment. Over all, their economic and financial position is more solid and stable than that of informal enterprises. In addition, the conditions of employment are considerably better in the formal sector. Being 'protected' by Labour Acts and trade unions, salaries of workers are considerably higher and more stable. Besides, their jobs are more secure and they enjoy privileges that workers in informal enterprises lack, like paid holidays, overtime payment regulations and insurance against accidents and illness.

According to the dualist approach, an important cause of the vulnerable economic position and worse working conditions in the informal sector is the weakly developed linkages of this sector with the formal sector. In this view, the formal and informal sectors are considered more or less autonomous entities, each with its own internal dynamics and mechanisms. Mutual linkages between them are, on the other hand, very rare. It has therefore, been recommended to Third World governments to support the informal sector by promoting and strengthening its 'upward and downward vertical exchange relations' with the modern enterprises and public institutions of the formal sector (ILO 1974; Moser 1978;

Smith 1988, 8; von Frieling 1989, p. 181; Gosses et al. 1989). By intensifying such relations, the informal sector's market for products or services would be enlarged and more modern technology would become at its disposal. If supported in this way, some informal enterprises would even become formal in the end. Moreover, the entering into contracts with informal sector enterprises would not be a sacrifice to most modern enterprises or public institutions, as informal sector enterprises are able to supply goods and services relatively cheaply. From the national perspective, stimulation of the informal sector would finally contribute significantly to a relief of the problem of urban unemployment. Compared with the formal sector, usually much less capital is needed to create one additional job in the informal sector (Sethuraman, 1981).

The structuralist approach has been developed as a critical answer to the dualist view. On the level of description of enterprises and employment, the majority of structuralist authors have never disagreed with most of the criteria used by their dualist 'opponents' to distinguish between the formal and the informal sector. They reject, however, the dualist premise that the formal and informal sectors are separated realms, each with their own exclusive characteristics, internal dynamics, and mechanisms of development. Instead, they emphasize the idea of mutual interrelations or linkages between the respective sectors or various segments in case more complex and fragmented structures are envisaged. Each segment derives its nature and dynamics mainly from its networks of mutual linkages with other segments within the urban production structure as a whole. Within the course of development of such networks, the component parts of the segmentation 'gradually have lost their individual identity and independence, so that we are now faced with one coherent whole; one system with its own character and dynamics' (Breman 1985, p. 53). It is stated, however, that this 'coherent whole' is neither static nor a-historical. On the contrary, it changes within the course of development (of global capitalism) and varies between different urban contexts.

Associated with the rejection of the idea of autonomous and mutually exclusive segments within Third World urban economic structures, the dualist concept 'sector' is also criticised by the structuralist approach. It has been replaced by the (marxist) concept 'mode of production'. This latter concept does not merely refer to internal characteristics of segments, but places these segments into the wider organisation of economic production and distribution as a whole. It does not only take into account the material and technical aspects of production in each of the segments, but also the social relations of production and the linkages between the segments. In addition, the linkages between segments, and concomitantly the economic characteristics of enterprises and conditions of employment, are considered to be modelled by the juridical-political *superstructure* of the (urban) society in question. Thus, the concept 'mode of production' surpasses the level of single enterprises.

Generally speaking, two types of modes of production have been distinguished: a capitalist mode, introduced from the First World capitalist and industrialised economies into the Third World, and an undefined number of domestic non- or pré-capitalist modes. The latter type however, has increasingly been transformed into a distorted version of the non-capitalist mode. Thus, both types of production modes have come into existence in one and the same historical process and they are two sides of the same coin. The informal sector is not traditional or lagging behind; both sectors are the product of capitalist penetration and contemporary technical modernisation (Santos 1979, p. 31).[1]

Within the 'coherent whole' of the urban economic structure, economic and political power is distributed unevenly over both types of production modes: the capitalist type has become dominant, while, as a consequence of their linkages with the capitalist, the non-capitalist types have lost their autonomy and have become subdued to the capital accumulation within the capitalist type (Forbes 1981, p. 109). In other words, the backwardness and impotence of the non-capitalist modes is preconditional to the development and progress of the capitalist mode. This dominance of the capitalist mode of production leans to some extent on the large, capitalist firms' political influence and power within the societal superstructure. Their power has put up the local juridical-political *élite* to create a framework of legislation, credit and tax policies, physical infrastructure etc., in favour of their capital accumulation and, as a reward for the rendered services, also in favour of the élite.

The dominance of large, capitalist or corporate type firms manifests itself in the transfer of value from the non-capitalist modes of production to the capitalist mode. This process, also considered the appropriation of surplus by the capitalist mode from the non-capitalist modes, is not just a matter of direct payments; it is characterised by a simultaneous occurrence of various exchange mechanisms. Lewin (1985, pp. 116-118) distinguishes three different types of such mechanisms of surplus appropriation: the implicit one, the indirect one and the direct one. Surplus is appropriated implicitly through the non-capitalist modes' function as a large and cheap reserve army of labour for capitalist firms and through the supply of cheap goods and services, produced by petty producers, to workers in the capitalist firms. In both ways, the non-capitalist modes of production exert a downward effect on the wages in capitalist firms. Indirect appropriation of surplus is realised by means of the market mechanism, supported by 'regulating functions' of the state. Large capitalist firms control the prices of supplies to and sales from non-capitalist enterprises and have an exclusive or near-exclusive entry to the consumer market of greatest purchasing-power. Finally, surplus is appropriated directly by means of subcontracting work to petty producers as to unload some risks (of slack seasons or production below quality standards etc.) and costs (for insurances or various kinds of inputs etc.) on these 'weaker shoulders'.

The mutual relations of dominance and dependency between the capitalist or corporate type firms on the one hand, and the semi- or non-capitalist petty type producers on the other, may both account for the latters persistence and their disappearance in the course of the development process. As has been pointed out in various studies on urban production relations in Latin America and SE Asia (Roberts 1978; McGee 1981, 1985; de Soto 1989), petty type producers usually are surviving in the meshes of the capitalist economic network as long as the capitalist enterprises are not interested in the poorer market segments served by these petty producers. Moreover, many petty type activities appear to fulfil important functions in 'subsidizing' the capitalist mode with cheap labour, raw materials, goods and services, thus facilitating its expansion in the national economy. Consequently, the forces of conservation in these petty type enterprises tend to domineer over the dissolution forces ensuing from corporate enterprise competition and substitution effects (cf. McGee 1981, 1985). Most petty type producers however, are only able to survive through rigorous self-exploitation of labour, reflected in long working days and very low incomes per working hour. Their somewhat paradoxical presence and persistence in especially the larger cities with a dynamic 'formal sector' economy therefore, should be explained by their subsistence on the circulation of money and goods generated by this formal sector, as well as by their intrinsic adaptability towards changing production conditions.

The structuralist approach agrees with the dualist observation of, over all, worse working conditions in what they label the non-capitalist modes of production. The large numbers of non-capitalist enterprises can continue their activities only by means of irregular working hours, extremely low and insecure incomes, and minimal levels of protection and certainty of employment. In addition to only describing these working conditions, the structuralists explain them as a consequence of the mechanisms of surplus appropriation. Changes for the better in the realm outside the capitalist firms and public institutions are hardly imaginable as long as the juridical-political superstructure supports the mechanisms of value transfer from the non-capitalist modes of production to the capitalist one. Among the variety of conditions of employment outside the capitalist sector, conditions are worst for what is called the 'marginal labour force' (Quijano Obregon, 1974), the 'stratum of the true self-employed' (Bromley & Gerry, 1979), or the 'residuum' (Lewin, 1985). People conducting those 'survival activities for the poorest' are not only more or less superfluous from the point of view of capitalist interest, and therefore expelled from any direct linkage with this mode of production, but they are also excluded from the more remunerative branches or enterprises in the non-capitalist modes of production.

Later on however, a debate arose within the structuralist approach on the marginal position of so-called self-employed workers in non-capitalist activities. In fact some types of petty producers seem to enjoy a less marginal position than many factory workers or even the lower employees in so-called 'formal' sector jobs (cf. Hagen Koo 1981; McGee 1981; Titus 1985). Especially the petty producers operating with skills and some working capital as intermediaries between the two modes, often enjoy higher incomes and a stronger position vis à vis pressures from the market or local authorities. Instead of being labelled as an impoverished sub-proletariat many of these petty producers might rather be qualified as proto-proletarians (McGee 1976, 1985), or even as lower middle class enterpreneurs. Considering the enormous differentiation in petty type activities and their different bargaining positions vis à vis the capitalist or 'formal' sector economy, it now seems untenable to view the petty producers as a homogeneous mass of marginalised workers who are more or less superfluous to the capitalist mode of production. As will be seen in the next sections this has important consequences for the structuring of the urban labour market and the conceptual models derived from it.

Views on the Segmentation of the Urban Economy

Thus far, the differences in theoretical points of view have been discussed between the dualist and the structuralist approach, including those on the role of the state in the development of the urban production and employment structures. These differences are clearly reflected in the way the two approaches deal with the main postulate in the general discussion on the Third World urban production and employment structures: its segmentation.

As has been said, dualist writers suppose the formal and the informal sector each to have its own exclusive characteristics. They have set themselves the task of listing these characteristics. Hence, dualism pretends to define the segmentation of the urban economic structure by means of sets of mutually exclusive characteristics that distinguish unequivocally between a formal and an informal sector.

The classical attempt to compose such a list of characteristics is made in above mentioned Kenya-report of the ILO of 1972. The informal sector would be characterised by ease of entry (for new enterprises) to the market, reliance on indigenous resources, family ownership of enterprises, small scale of operation, the use of labour-intensive and adapted technology, skills (of workers) acquired outside the formal school system and by unregulated and competitive markets. The characteristics of the formal sector would be the reverse of these: difficult entry, frequent reliance on overseas resources, corporate ownership, large scale operation, the use of capital-intensive and often imported technology, formally acquired skills (often expatriate), and protected markets (through tariffs, quotas and trade licenses) (ILO 1972, p. 6).

In the structuralist approach, the segmentation of the urban economy is interpreted as a continuum. In that continuum the capitalist mode of production is still considered one separate and undivided entity of large, registered capitalist firms with an organised and protected labour force earning regular salaries and enjoying secure employment. The variety of economic activities outside this mode, however, is so enormously fragmented that every attempt to classify them into a fixed number of distinct classes means an inadequate simplification of reality. In fact, the entirety of activities outside the capitalist mode of production has a crumbled structure in which borderlines between composite parts are arbitrary and difficult to locate (Breman 1985, p. 51; Gilbert & Gugler 1984, p. 77). This crumbled structure reflects the large variety of forms and mechanisms in which these activities (enterprises) are mobilised in the process of capital accumulation of the capitalist firms. Besides, some activities seem to be superfluous from the point of view of capitalist interest and, consequently, have no direct linkages whatsoever with that dominant mode of production.

Problems in Applying These Views to the Empirical World

At first sight, sets of criteria like those of the ILO Kenya-report to define the formal and the informal sector appear to be well-considered. Nevertheless, as operationalisations of the dualist approach they suffer from some serious difficulties. First, unequivocal classification of each enterprise into either the one or the other sector has often appeared impossible in real-world situations: many, and in some cases even the majority of enterprises are characterised by a mixture of the selected formal and informal characteristics. To deal with this, most of the applications of such lists to the reality of urban centres have brought together into the informal sector 'everything' that does not belong to the registered and easily to identify formal sector. This 'solution' has provoked the criticism that the informal sector 'defined' as such is nothing but a residual category and that no detailed insight is given into the large variety of activities outside the formal sector.

Apart from this shortcoming, dualist studies also suffer from a lack of consensus in three different fields: the units of analysis, the criteria to distinguish sectors and the names of sectors. Concerning the units of analysis, the two that have been used most frequently are enterprises and workers/working conditions. Occasionally, however, income strata, inmigrant populations, or even types of neighbourhoods are selected (Moser, 1978; Bromley, 1978; Hoenderdos, 1982). In addition, various criteria to distinguish the two sectors have been used. In studies in which enterprises are classified, criteria stretch from registration at the Municipality Office to the number of workers or the value added per worker. In case of workers being the unit of analysis, the protection of their working conditions by law and trade-unions or the certainty of employment and income are often considered the main criteria. Finally, although the dichotomy of formal versus informal sector is the most common one, some other names have to be mentioned as well. Examples are the protected versus unprotected sector of Mazumdar (van Dijk 1978, p. 12) and the high profit & high wage international oligopolistic versus low profit & low wage competitive capitalist sector of Brown (Moser 1978, pp. 1052-1053). Over all, such a large variety of entries make dualist studies hardly comparable and have led to a complete confusion about what is actually meant by the informal sector (Moser 1978, p. 1051; Smith 1988, pp. 37-41).

Most of the above mentioned criticism at the dualist' attempt to segment urban economic structures is levelled by structuralist authors. Their own idea of a continuum of segments without fixed boundary lines on the other hand has raised insurmountable difficulties of operationalisation. In many case studies, this problem has been by-passed by studying only one (tiny) fraction of the entire urban economy outside the capitalist mode of production, in some cases no more than one type of workshop or subcontractor. Examples are two case studies on the relations between capitalist institutions or enterprises and subcontractors in Cali, Colombia, applying respectively to the lottery business (Gerry, 1985) and the production of recycling paper (Birkbeck, 1979), and a study on pedlars and *trishaw* riders in Ujung Pandang, Indonesia (Forbes, 1981). In addition, the lack of consensus with regard to the unit of analysis, the criteria to distinguish between segments and the names of these segments can also be found among structuralist studies. Examples of two studies that differ in all these three fields are those by Bromley & Gerry (1979) and by Titus et al. (1986). Bromley & Gerry classified forms of 'working relationships' into five classes, ranging from stable wage-work to true self-employment, whereas Titus et al. classified enterprises into a continuum between petty commodity and corporate type enterprises.

Applications to the Urban Production Structure

Some adherents to the dualist approach accepted the structuralist criticism on the idea of a bipartition of urban production and employment structures. Even at the ILO it was admitted that, in some cases, it may be necessary to distinguish more than two sectors (Sethuraman, 1981). Besides, it was also admitted that each city is unique and that it is unwise to strive for universally valid lists of criteria. The structuralist authors, in their turn, have not succeeded yet in solving the difficulties of operationalising their notion of a continuum of segments without clear borderlines. Consequently, in spite of their rather different points of view, adherents to both approaches have designed tripartitions, quadripartitions etc. in order to analyse urban production and employment structures.

In the next sections, some examples of such segmentations will be presented rather extensively. The specific selection of these examples emphasises the broad variety that has developed within the analysis of urban production and employment structures. Awareness of this variety is the basis of a well-considered theoretical and empirical framework to examine the particular case of small towns.

An early example of a production structure model that includes more than two segments is the one designed by van Dijk (1980). The used unit of analysis is the enterprise and Van Dijk's theoretical position is close to the ILO approach. In his study to the secondary sector in the West-African capitals of Dakar (Senegal) and Ouagadougou (Burkina Fasso), he distinguished among the formal modern industrial sector, the small scale industrial sector and the informal sector. Because of its considerable internal variety, the latter sector was further subdivided into a more and a less developed part. He constructed the tripartition in two stages. First, he selected three relevant criteria to isolate the formal sector by posing three questions: do enterprises possess a juridical statute, do they pay the legal minimum salary to the employees, and had they registered their employees at the *Caisse de Sécurité Sociale*? The criteria were borrowed from the Labour Acts that were in force in most countries in French speaking West Africa and that obliged enterprises to meet these three criteria. Next, the enterprises that did not meet the criteria, i.e. the non-formal ones, were further subdivided into two sectors. To do so, the above mentioned requirements inserted in the Labour Acts were supplemented with some new criteria. The small scale industrial sectors were distinguished on the basis of a legal status to the Treasury, the obligation to obtain a juridical statute, and being considered for governmental support. In addition, employees of such enterprises would be registered at the *Caisse*, which, by the way, did not mean a guarantee that they would receive the legal minimum salary. Finally, enterprises in the informal sector had no statute, did not pay the legal minimum salary and had not registered their workers. In conclusion, van Dijk based the already mentioned further subdivision of the informal sector into a more and a less developed part on the criteria 'use of expensive equipment' and 'non-family labour in service'.

A second example of a model that is designed to classify enterprises is the already mentioned one by Titus et al. (1986). Unlike van Dijk, Titus et al. sympathised explicitly with the structuralist view of urban production structures (Titus et al. 1986, p. 254). In order to segmentize urban production structures, two dimensions were distinguished. The first dimension concerns the capitalist nature of enterprises. According to the structuralist view, this dimension is thought to be the outcome of the confrontation between modes of production on the level of individual enterprises. To measure this dimension, a specific set of characteristics was distinguished; the structuring characteristics. This set of six types of characteristics includes the capital intensity of enterprises, their access to formal credit, and several indicators on their social relations of production. The degree to which an enterprise is capitalist or non-capitalist depends on the scores of these structuring characteristics. Besides the capitalist dimension, another dimension, one that is not perse associated with the confrontation of modes of production, was also considered relevant from the perspective of segmentation. This second dimension is 'related to the functional appearance of enterprises' (Andriessen & Van de Broek 1988, p. 9). The type of characteristics to measure this dimension was labelled form characteristics. These form characteristics involve items like the size of an enterprise and the use that is made of modern equipment i.e. criteria determining the corporate (large scale, modern equipment etc.) or petty commodity character (small scale, no modern equipment etc.) of enterprises.

Thus far, the model of Titus et al. has been applied to four small urban centres in Central Java (De Jong & Ligthart 1986; De Jong & Steenbergen 1987; Andriessen & Van de Broek 1988; Verhoog & Van Rijn 1989). In these case studies, the structuring and form characteristics were first operationalised into respectively six and four directly measurable variables.

In the next step, each of the selected structuring and form characteristics was reduced to a dichotomous variable: the value '0' as the equivalent of non-capitalist and petty commodity and '1' of capitalist and corporate. The sum of the numbers of capitalist characteristics of a surveyed enterprise or institution equals its score on the first dimension and the sum of the number of corporate-characteristics its score on the second dimension. These two scores were then put into a two-dimensional matrix. Thus, each cell of this matrix represents a particular numerical combination of corporate and capitalist characteristics. Finally, the matrix may be divided into two segments by the drawing of a diagonal (table 2.1), separating the predominanty corporate enterprises from the petty type enterprises.

In the four applications thus far, the model of Titus et al. has yielded only a new bipartition of the urban production structure at a higher level of generalisation. At a lower level, however, it should also be possible to consider each cell of the matrix to represent a *distinct* type of enterprises. Then, six structuring- and four form-characteristics yield no less than thirty-five types of enterprises. Such a large number is illustrative of the fragmented and crumbled character of the urban production structure as advocated by the structuralist approach.[2] The most promising application of the model however, would be the grouping together of clusters of cells into a smaller number of segments. The precise grouping together may depend on the *scattergram* of enterprises within the matrix. In this way, segments should not be defined a-priori, but chosen freely depending on the particular empirical situation in the town or city the model is applied to, and without losing the information about the internal variety of each of the segments.

Table 2.1: Enterprise matrix scoring on form and structuring characteristics

Form characteristics	Structuring characteristics						
4				I	W	I	
3			W		TI		
2		P	TP	T	TW	TT	
1	P	PM	PP	P			
0	M	M	MP				
number of characteristics	0	1	2	3	4	5	6

Enterprise types:
I = corporate industry
M = market trader
P = petty producer/trader
T = toko (shop) trader
W = wholesale trader

Applications to the Urban Labour Market

The earlier conceptualisations of the urban labour market, assuming a dichotomization between formal and informal types of employment, were criticised for both their theoretical weaknesses and their operational shortcomings. Much employment in so-called formal sector jobs did not appear as permanent, as formalised and fully employed as envisaged in the original conception. Moreover, entrance to these types of jobs often appeared to be determined by informal patron-client relationships and other 'traditional' types of credentialism (ethnicity, bribary) instead of formal criteria like skills and experience (Fields 1990). Finally, recent developments like the process of economic globalization and the structural adjustment programmes seem to have contributed to a further informalization and 'casualisation' of formal sector work, thus blurring the old distinction between the formal and informal sections of the urban labour market (ILO 1993).

A first attempt to overcome the shortcomings of a dichotomized labour market consisting of two opposite sectors with mutually exclusive characteristics, has been made by Friedmann & Sullivan (1974). Starting from the organisational characteristics of the main types of enterprise activity in which the urban labour force participates, the authors distinguish a corporate enterprise sector, a family enterprise sector and an individual enterprise sector. In each sector there are possibilities for achieving well or less integrated positions. The first sector, consisting of large scale corporate type enterprises, the public sector and large family firms, offers the most stable employment conditions and best paid jobs. The second sector consists of smaller family owned firms, employing the entrepreneur and a few workers in trade, services or small scale industries. Much of this labour is employed as (un)paid family labour in relative secure, but lowly paid jobs. The third sector employs only own account workers and independent producers with very limited skills and means of production and consequently offers the least income security. This sector is seen either as a waiting room or as a dead-end alley for those who are still aspiring for employment in the corporate sector, or those who ultimately have failed to achieve their aim. Employment in the three sectors thus is arranged in an ascending scale of labour productivity and social status, so that the average level of integration in each sector is higher than in the previous sector.

The last example discussed here is the model of Bromley & Gerry (1979). This is a theoretical model, not based on field research in one specific urban centre. The model deals with 'forms of working relationships' instead of enterprises. Bromley & Gerry started from the idea of a continuum of such relationships. Stable wage work in capitalist firms or government offices is at one extreme. This extreme contrasts with a broad range of working relationships they put under the collective header 'casual work'. This casual work was then, however, roughly divided into four broad and occasionally overlapping ideal-type categories: short-term wage work, disguised wage work, dependent work and true self-employment. These four categories differ from stable wage work to an increasing extent. Hence, what was meant to be a continuum, in practice became a quintipartition of working relationships.

Regarding the four types of 'casual work', short-term wage workers are contracted by a capitalist firm for a specific term or task. Although such workers are not assured of continuity of employment, they are employees of the firm in a legal sense. This is not true for distinct wage-workers, who are mostly outside-workers or subcontractors. They are often provided with equipment, raw materials or other inputs, but lack the protection and other advantages offered by the status of short-term wage-workers, like a fixed salary, paid holidays, and insurance against accidents. Dependent workers have no wage relation whatsoever with a large capitalist firm. They work for their own account and have to purchase or rent equipment, raw materials etc. from large firms themselves. Because such firms are often monopolists or oligopolists, the dependent workers pay high prices or rents without many possibilities to switch to another supplier. Such a shift is even impossible when their disadvantaged position has led to an indebtment relation. True self-employed, finally, have no relation at all with the capitalist sector. They lack secure supplies of inputs and their employment and income are most insecure of all four categories of casual work.

Both the labour market models of Friedmann & Sullivan (1974) and Bromley & Gerry (1979) suffer from the implicit assumption that the urban labour market is hierarchically organized as a continuum of gradually improving job opportunities ranging from casual, unskilled and self employed work in the pre-dominantly non-capitalistic enterprise sectors to permanent, skilled and salaried jobs in the capitalistic or corporate sectors. This view has been challenged by authors like Breman (1977), Forbes (1981) and McGee (1981), who stress the fragmented nature of the urban labour market due to the unequal impact of capitalist transformation processes and institutional factors. The resulting structure of the urban labour market shows a multitude of more or less separate compartments or niches, which are protected by different conditions of entry and credentialism. According to these authors empirical evidence shows that marginal positions can be found in all sectors of the urban economy, including the capitalistic or corporate enterprise sector. Thus, casual work is encountered in both the capitalistic and non capitalistic enterprises. Moreover, some types of casual workers in the capitalist mode need not be better off than casual workers in the non-capitalist mode, whereas on the other hand some self-employed workers in the non-capitalist mode may be better off than many permanent wage workers in the capitalist mode. Labour absorption and labour mobility for the unskilled masses however, usually remains confined to the lower reaches of the various labour market segments.

Although Breman (1980) and McGee (1981) made important points with respect to the fragmented structure of the urban labour market, they refrained from the construction of an operational model which could be tested empirically. Such a model has been devised by Titus (1985) in his dissertation study on migrant integration and demographic changes in Jakarta, and since then the model has been used by several authors in the SOREGIO 'small towns' research programme (cf. Van der Post 1988; van der Wouden 1991; Harts-Broekhuis & de Jong 1993; Romein 1995).

Actually, the structuralist model is based on a combination of Friedmann & Sullivan's enterprise type model and Aldunate's integration level model. In line with the latter Titus (1985) distinguished a high, a medium and a low level of integration in each enterprise sector, defined by criteria like income, job security, skills, employment status and involvement in the maintenance of the institutional order. Labour participation in the corporate, family and individual enterprise sectors thus is classified into three levels of integration or nine positions, suggesting that in each sector labour can be absorbed in a core position (H-level), an intermediate position (M-level) or a marginal position (L-level). It should be noticed however, that due to differences in accessibility, income levels and labour relations, the levels of integration in each sector are not directly comparable with corresponding levels in the other sectors. By placing the sectors and their respective positions/levels next to each other - instead of in a hierarchical order - the a priori assumption of a subordination of the individual sector's integration levels to those of the others is avoided.

Table 2.2: The adjusted model of the urban labourmarket

Level of Integration	Corporate Type Sector	Family Type Sector	Individual Type Sector
High	C-1	F-1	I-1
Medium	C-2	F-2	I-2
Low	C-3	F-3	I-3

The model however, enables us to draw *a posteriori* conclusions on the relative position of the sectoral integration levels vis à vis their average income and employment conditions. Remarkably enough, Titus' (1985) research revealed a new dichotomy between labour absorbed in positions situated to the left and to the right of the diagonal running from the lowest position in the C-sector to the highest position in the I-sector, thereby suggesting that especially labour employed in the lower F- and I-sector positions had been marginally integrated into the urban economy.

Evidence from Small Town Studies

Until recently research on the small town production and employment structures has been rather scanty and inconclusive, frequently showing more interest in the towns productive functions towards their hinterlands than in their intrinsic structural characteristics[3] (cf. Hardoy & Satterthwaite 1986). On the other hand, most studies on urban production structures and labour absorption have been carried out in large cities with relatively dynamic economic structures and high rates of in-migration, thus displaying quite different production and employment conditions. If we confine our review to small and intermediate towns with less than a 100,000 inhabitants, most of these towns show a strong relationship with their rural hinterlands through the predominance of their servicing functions. This implies that next to large distributive and collecting trade sectors and some agricultural processing, these towns often have a sizeable public sector. Especially the district capitals show large administrative sectors, which are the main providers of formal sector employment, as well as a main source of financing. The small secondary sector in these towns usually shows little dynamism, as it consists mainly of individual and household enterprises using simple techniques and serving local needs only. In some cases however, like in company towns, corporate enterprises may be ostentatiously present.

Most authors agree on the fact that small towns usually have a smaller formal or corporate type economy than the large cities. The main reason for this being the relative high threshold for viable corporate enterprise activities in terms of local demand and productive facilities. Many towns simply are too small, have too weakly developed infrastructural facilities and a hinterland population which is too poor, to be attractive for this type of enterprises (cf. Davila & Satterthwaite 1987). Consequently, informal or petty type activities constitute the overwhelming majority in the small town's economic structure, while the formal or corporate activities are strongly dominated by public and commercial services. Another aspect of this lopsided structure is the relative stagnation of the town's informal sector economy which lacks the support of a dynamic corporate sector generating a sufficient circulation of money and goods upon which it may subsist (cf. Pedersen 1990).

The small town's employment structure usually closely reflects the production structure, showing a predominance of civil servants, shopkeepers and large traders in the higher echelons of the formal labour market, which are protected by less accessible credentials and financial requirements, whereas the mass of own account workers and petty producers is found in the more accessible lower end jobs in the informal labour market. Moreover, many of these lower end jobs are occupied by (un)paid family workers like domestic servants, shop assistants, apprentices, etc. The small town labour market therefore, may still display strong dichotomous characteristics although pre-capitalistic labour relations are not exclusively attached to the lower end jobs and capitalist features have pervaded all segments of the labour market (cf. Harris 1982; Van Lindert en Verkoren, 1991).

Occupational mobility between the two realms of the small labour market usually remains limited because of the required credentials and the saturated employment conditions in the upper market levels. Few migrants from the hinterland areas therefore, are attracted by the towns formal sector opportunities, while on the other hand the overcrowded and stagnant conditions of the towns informal sector economy explains for the lack of interest of even the unskilled migrants from these areas. It should be kept in mind however, that there may be important exceptions to this highly generalized picture of the small town's production and employment conditions. In so-called company towns, arising from the establishment of a mining company or 'maquiladora' industries for example, the corporate sector may be very dominant. Moreover, small towns in resource frontier areas usually fulfil vital functions for a wide area and seem to attract many migrants, thus displaying more diversified production and employment structures than might be expected according to their recent establishment and relative small size. Finally, the small town's production and employment structures appear to be strongly related to both the rural hinterland conditions and national development policies. Densely populated and commercialized rural hinterlands offer better opportunities for developing the towns servicing functions and diversifying its economic structure than sparsely populated, subsistence farming areas. Similarly, national development policies stimulating rural development will be more favourable to small town development than urban biased policies like import substitution industrialisation.

Conclusion

A debate on the theme of Third World urban employment and production structures started in the 1960s with a straight dualist point of view. After two decades of structuralist criticism however, this thematic field has crumbled into a heterogeneous landscape of gradually different theoretical positions and empirical partitions. Hence, the well-ordered division into two opposing theoretical approaches that has been made is a highly artificial one. There is no more easy choice to be made between the one or the other theoretical blueprint and associated fixed set of instruments. This is far from regrettable however, because it creates the opportunity for any investigator of the production and employment structures of small towns to make a flexible choice out of a wide range of theoretical premises and empirical instruments of analysis, according to the central research question and the specific context of the towns concerned.

Although theoretical premises and empirical instruments need to be selected flexibly, two decades of thinking on Third World urban production and employment have taught us some important lessons that should be kept in mind profoundly in conducting research in small towns. First of all, research should focus on analyzing the characteristics of the small town production structure and production relations, by using the enterprise with its mode of production characteristics as the basic unit of analysis and paying special attention to the functional relations between the various types of enterprise (leading to processes of conservation or dissolution). Secondly, considerable importance should be attached to the analysis of the small town employment structure and employment conditions, by using the household as a basic unit of analysis and paying special attention to the structuralist characteristics of the urban labour market, the absorptive capacity of its various segments and the type of labour relations prevailing in each of them. Thirdly, special attention should be paid to the functional relations with the town's hinterland economy and society, as well as with higher order centres, as these may be crucial in shaping the town's production and employment structures through mechanisms of exchange and surplus transfer, competition, labour migration, etc..

An additional important factor is the role played by the state or government in the development of the urban production and employment structures. The construction of physical infrastructure, the economic behaviour of the government -either as a customer or a producer- and the creation of an institutional framework, i.e. the whole fabric of legislation, regulations, subsidies etc., are of vital importance to the small town's structure and functioning. Moreover, many characteristics of the urban production and employment structures can only be understood against the background of the economic, social and cultural development of the society as a whole. In order to gain insight into the production and employment structures of the specific case of small towns, one has to be aware also of relevant processes and mechanisms on the level of the national economy and society. After being identified, such processes and mechanisms need to be directive, in some way or another, for the selection and formulation of the research questions and the set of directly measurable variables. Unfortunately, thus far empirical research has not been a great help: the structuralist approach has stressed the 'social embedding' of research to Third World urban production and employment but empirical studies are still scarce.

Finally, Third World urban production and employment structures even ought to be placed within the context of the (capitalist) new international division of labour (NIDL). Over all, the traditional role of the Third World within the international division of labour, as a supplier of raw materials, has been modified since the early 1970s. The *new* international and division of labour refers to a rapid process of commercialisation and industrialisation in the Third World. Direct foreign investments (DFI) in developing countries, very much concentrated in urban areas, are a very important determinant of this process (Chandra 1992). In the urban areas where this has occurred, the NIDL is an important framework of reference to examine their production and employment structures.

Thus, important evidence has been found for example on the paralyzing effects of structural adjustment programmes and subsequent government budget cuttings in some corporate enterprise activities, i.e. especially among state enterprises and private enterprises depending on government expenditures (Moser 1993). On the other hand, the process of globalization and related foreign investments may even engender an expansion of informal sector activities through the increase of putting out systems and sub-contracting relations which often are favoured for reasons of international competition. Correspondingly, processes of economic globalization may strongly affect the urban labour market through a progressive 'casualisation' of the labour relations, both in the formal and informal types of employment. However, as DFI are distributed very unevenly among Third World cities and the small town usually is one of the least favoured areas for the establishment of any factories or other corporate type activities (management, Research & Development etc.) by

Transnational Corporations, it is questionable whether these processes are also relevant for small towns.

Notes

1. Some authors hold the view that, just because petty commodity production is an integral part of the capitalist mode of production, albeit a marginal and subordinated part, it cannot be a separate mode of production. In stead of a mode of production, it is called a 'form of production' (Moser 1978; Forbes 1981) or an 'economy' (Castells & Portes 1989).
2. A disadvantage is the large sample that is necesarry if one wants to draw conclusions on each of these types of enterprises.
3. Intrinsic structural characteristics are conceived here as characteristics directly deriving from the small towns' functions as administrative and servicing centres and from their hinterland characteristics.

CHAPTER THREE
Town and Hinterland in Central Mali: The Case of Mopti

Ali de Jong & Annelet Broekhuis

This chapter focuses on two topics: the different economic relationships between a medium sized town in Mali and its surroundings and the production structure of this town. These topics will be treated separately. The town of Mopti[1] and its hinterland are located in the Sahelian zone of Mali, where the population is confronted not only with economic problems, but also with ecological degradation. These problems are far from unique. In the entire Sahelian zone, and even in the majority of the African countries, the economy and the environment give cause for great concern.

The research on which this chapter is based took place during the period 1983-1991[2], a period in which the population particularly suffered from harsh economic conditions and important adjustments in the development policy of the central government. During the seventies and eighties extreme annual fluctuations were observed in the agricultural production of the Mopti Region, and consequently large groups of the population were faced with food deficits. Issues such as insecurity of subsistence and survival were found to be more prominent than changes linked with economic or agricultural development. This lack of economic development was reflected in a relatively modest annual population growth rate of 1.07 percent for the period between the censuses of 1976 and 1987; a figure lower than the national average and also lower than the region's past population growth.

The Regional Setting

The Mopti Region, with an area of 88,752 square kilometres more than twice the size of the Netherlands, is located between long. 2°60' and 4°30' W. and lat. 14° and 15°50' N. Its location in the semi-arid tropics, the transition area between the humid tropics in the south and the desert in the north, determines its physical characteristics and its agricultural production potential. Another physical determinant is the so-called Niger Inner Delta, a vast plain of about 30,000 square kilometres which each year is flooded to some degree by the River Niger, the Bani and their tributaries. The presence of this flood plain enables a larger variety of modes of subsistence than is possible in other areas within the semi-arid tropics. Both precipitation - the meteorological station of Mopti registered an annual average over 400 millimetres in the period 1979-1990 - and inundation are characterised by great annual and seasonal variability.

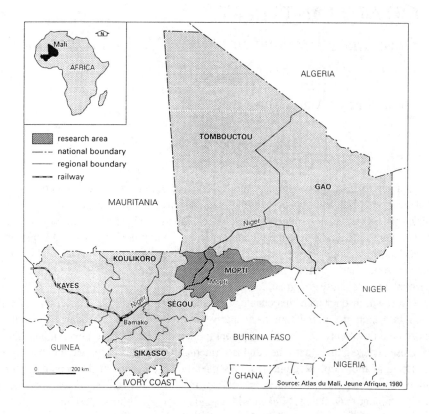

Figure 3.1: The Region of Mopti in Mali

This variability makes agriculture a risky activity and causes the production in animal husbandry and fishing to fluctuate sharply over the years: for instance an analysis of the areal photographs taken between March 1981 and May 1987 illustrated that the number of cattle in the Delta decreased by almost forty percent (RIM 1987), while the number of small ruminants increased. The average size of fish catches in the mid-eighties was only slightly more than half of that in the seventies. Despite these physical hazards at the time of our research, primary production remained the basis of the regional economy; trade and secondary production obtained their products and raw materials mainly from agriculture, animal husbandry and fishing, and most of the population (about 90%) was dependent on primary production activities. The relative importance of the various activities was difficult to assess because the majority of the rural households combined several activities with a varying output and labour input over time. Even among the urban population, agriculture and animal husbandry were important activities and sources of income.

The low and variable precipitation, the high degree of evapotranspiration, the short growth season and the low natural fertility of the soils restrict the agricultural possibilities.

Millet and sorghum, crops which are relatively drought resistant and which make limited demands on the soil, are the most important rain-fed crops. They are grown in the eastern part of the region. In the western part, the regularly inundated delta area, rice is grown. Average yields are very low. Millet and sorghum yields vary around 500 and 600 kilograms per hectare respectively. The avarage yield of the traditional red rice varieties cultivated for centuries on the flood plain of the Inner Delta was calculated at 780 kilograms per sown hectare for the period 1974-1991, but with annual variations between 200 and 1,600 kilograms. The yields of the modern wet rice varieties cultivated in the constructed polders did not exceed those of the traditional varieties: the average yield until 1991 amounted to 670 kilograms per sown hectare. An analysis of the evolution of the yields even points to a decline caused by decreasing soil fertility. Since the eighties rice is also cultivated in a growing number of small village polders which can be irrigated all the year round and which make possible high production levels with up to 8000 kilograms per hectare.

Despite the fact that agriculture is considered to be by far the most important form of production in terms of employment, the Mopti Region is not known for its agricultural products. Both fishery and animal husbandry are regarded as the typical activities of the region (RIM 1987). This is partly explained by the fact that these two types of primary production supply the national market and also certain foreign markets, while the total cereal production in the Mopti Region has become increasingly inadequate to even meet the demand of its own population.

Not only primary production activities but also such urban-based activities like trade and transport are characterised by a pronounced seasonality, which means that the urban economy of Mopti town shows peak periods as well. This urban economy has adapted itself to changing demographic and economic conditions. From its creation at the beginning of this century, Mopti has seen periods of rapid population growth. The impetus to this population growth came from three stimulating factors: geographical situation, external interventions and trade. Mopti's location, at the confluence of the Niger and the Bani, turned out to be a very favourable one for fishing and trade and for colonial strategic aims. The construction of a hard-surfaced dyke which connects Mopti with the national road, from the capital Bamako to the north, opened up new outlets in the southern regions for its fish and cattle (southern Mali, Ivory Coast, Ghana etc.) and also made Mopti a trans-shipment point in land and water transportation routes. The Niger River connects Mopti to the west to Koulikoro - the harbour of Bamako and the terminus of the Dakar-Kayes-Bamako-Koulikoro railway - and to Gao and Tombouctou to the east. Investments in the national road transport network from 1960 onwards facilitated the relations with the southern regions and the capital, and connected Mopti to other seaports such as Abidjan and Lome. The national road from Bamako to the north, which was only completed in 1986, also linked the town with Gao.

After independence from France in 1960, Mopti was designated as the regional capital of the fifth region, the Mopti Region. This led to a spectacular growth of the government apparatus and to an influx of civil servants and their families with comparatively strong purchasing power. The establishment of a number of state agencies and development organisations in the seventies contributed especially to the increase of the government sector.

In the early nineties Mopti - with over 74,000 inhabitants - ranks third after the national capital Bamako and Ségou in terms of population size. Mopti is the most important town in the three northern regions of Mali. It outnumbers the two other regional capitals Tombouctou and Gao in size and economic importance, and in its own region the town dominates the urban hierarchy. The famous old town of Djenné, which ranks second has a population of only 14,000.

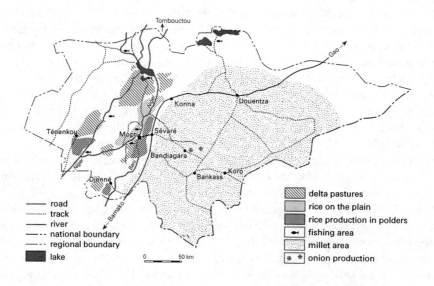

Figure 3.2: The Mopti Region: infrastructure, towns and rural production

Mopti and its Hinterland: Urban-Rural Relations Examined

In the regional planning literature of the seventies rural centres were seen as important agents of economic growth for their rural hinterlands. Some authors stressed the market functions of these centres (Mosher 1969; Dewar *et al*. 1986), others added the service functions (Van Dusseldorp 1971; ESCAP 1979). Rather generalised blueprints were elaborated with very little reference to population density, production levels, commercialisation of the production and purchasing power of the population. In the eighties Rondinelli (1983) emphasised the developmental role of secondary cities. He listed many functions which such cities in developing countries could have for their surrounding areas.[3] According to Rondinelli these cities provide the following economic services and facilities: decentralised public services in the areas of administration, health care, education; collection and distribution facilities and markets for agricultural products; processing industries; transport and communication facilities and non-agricultural activities for rural inhabitants.

Not everyone agreed with this positive approach of the functions of rural centres. Schatzberg (1979) even called small urban centres in Africa 'islands of privilege' referring to their general weak economic structure, a heritage of the colonial past, and a predominant civil service which does little to contribute to the development of the surrounding areas.

Titus (1986, 1991) tried to go beyond the seeming opposition between those authors, the majority, who see urban centres as (possible) vehicles of development, and those who see them as impeding rural development. He stressed the importance of the capitalistic and pre-capitalistic characteristics of the production structure in town and hinterland. According to him, depending on the degree of symbiosis or competition between enterprises and on the connected mechanisms of surplus transfer between rural and urban areas, urban-rural relations may restrict or stimulate development.

Mopti is a secondary town in the bureaucratic and planning organisation of Mali, a position shared with the capitals of the other six regions. All these secondary towns are far behind the national capital, a primate city, which has a much larger population and many more and specialised functions.

Being a regional capital, Mopti is the location of certain public services in the fields of administration, tax, education, health care, post and telecommunication, forestry and environment etc. In Mali these services are organised according to hierarchical principles: the top level is located in the national capital, the second level in the secondary towns, and the third and, if any, the fourth level in the centres of the districts and sub-districts.

To any visitor of Mopti it is clear that the city is not just an administrative centre for its hinterland. Its busy harbour, the active trade in the streets and markets in all kind of goods, the activities on the '*gare-routières*', the numerous truck movements carrying goods in and out and the many small shops point to Mopti's many other functions. Its favourable location as described above, makes the town an important node in transport, storage and trade networks. Apart from its relatively good access to the outside world, Mopti is surrounded by a diversified countryside with many more or less specialised forms of production. Mopti has become the locus of trade between local and (sub)national levels.

During the seventies and eighties some important development efforts in the Mopti region were oriented to agriculture, animal husbandry and fishery. The headquarters of the organisations concerned were located in Mopti. Development efforts were not limited to rural activities. Industrial development also was promoted: in Mopti several agro-based industries were created.

Many studies on relations between an urban centre and its surroundings presuppose (regional) economic growth. More recently and especially in studies dealing with Africa, there has been an interest in urban-rural interaction under conditions of economic stagnation and decline. The Mopti Region has undergone a period of economic prosperity in which rural production increased, urban-rural interaction intensified, and urban activities extended. The drought periods in the seventies and again in the eighties adversely affected the favourable developments. A period of economic stagnation set in. The relations between town and hinterland undoubtedly were influenced by these worsening conditions.

In the following sub-section we will look at the characteristics of the economic[3] urban-rural relations between Mopti and its hinterland, the eventual surplus transfer and at the effects of economic stagnation on these relations.

Urban Functions: Marketing, Processing and Service Provision
Right from the beginning of its existence, Mopti has been a major trade centre (Gallais 1967). Collecting trade in regional produce from the hinterland always has been an important function. During the research Mopti was found to be the most important fish market in Mali. It served a wide area, far beyond the region's borders. The catch from downstream and upstream fishing areas was collected - smoked, dried or fresh - and was distributed on the local and regional markets, or transported to the larger cities in Mali and even in Ghana, Ivory Coast and Burkina Faso.

The second most important trade product was grain. Millet from the eastern part of the Mopti Region was bought up by traders at the local markets after the harvest and stored and distributed in Mopti. Part of the millet was sold to other parts of the Mopti Region which were specialised in other products (the delta where fishing and rice growing and the north where cattle herding were more dominant activities), and part to other regions. Since centuries, rice was produced on the Delta plain, primarily for self-consumption. The creation of large polders and subsequently of many small village polders where rice was produced for the market added to the importance of the trade in rice.

Livestock was the third product of Mopti's hinterland. Cattle, sheep and goats were collected by way of a network of markets into the northern region of Gao. From the regional end-markets Konna and Fatoma (near Mopti), situated in Mopti District, the animals were exported outside the region to Ivory Coast and to the Malian towns in the south. The volume of the onion trade to Mopti from the Dogon plateau, and mats, earthen pots or woolen blankets from specialised villages was far less important than the trade in the three first-mentioned products.

The volume of collected products from the surrounding areas to supply the urban population of Mopti was small when compared to the volume bought up and sold to areas outside the Mopti Region. This can be explained by a number of factors. On the one side there was the fact that the purchasing power of Mopti's population was very low. So Mopti constitutes a relatively small outlet. The market for 'luxury' products such as fish and meat was small. On the other side there are: the presence of a certain group of traders in Mopti who had access to sufficient capital to wholesale and transport products; the favourable location of Mopti; and the presence of a number of government organisations in Mopti which during a certain period were entrusted with the buying up of cereals.

Distributive trade of goods from outside the Region to its local and regional populations appeared to be another function of Mopti. These imported goods may be specialised products from higher order centres which will be passed on to centres of a lower order in the hinterland. Indeed industrial products such as fuel, building materials, spare parts, tools, textiles, drinks and cigarettes were imported. Of more importance were imports of food: cereals, vegetables and fruits. The major part of the food was imported from overseas and consisted of cereals, often relief goods, but also commercially imported rice. The destination of these imported goods were Mopti town, its hinterland, and also the northern regions of

Tombouctou and Gao. Mopti fulfilled an important transit function for these northern regions.

According to Rondinelli (1983), the processing of agricultural produce is an important activity in secondary cities. Operating on a larger scale, the processing industry of the secondary cities would be able to produce at lower costs than the traditional small enterprises.

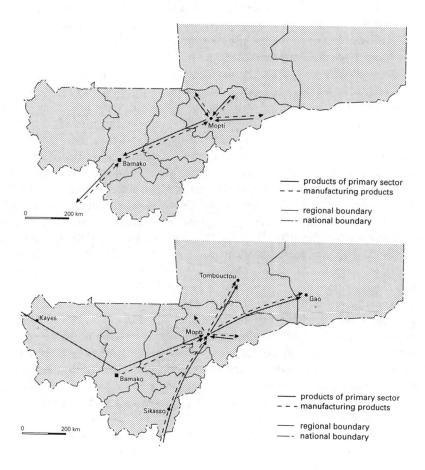

Figure 3.3: Theoretical and actual flows of products from and to the Mopti Region

In Mopti crafts were practised in many small enterprises. Most of this production was found to be destined for the urban market, with a small part of it for the rural markets. However, rarely did these crafts concern the processing of agrarian produce of the surroundings: there were some small grain mills, some leather working, some charcoal burning, and nothing more. The other crafts like weaving, tailoring, metalworking, shipbuilding etc. worked with imported raw materials. The two more modern processing bakeries in Mopti also processed imported meal.

Processing of collected produce from the surroundings took place in rather modern, relatively capital-intensive industrial enterprises. These enterprises were small according to international norms, employing ten to forty people, but were much larger than the small craft workshops. Fish was processed in the fish factory of the *Opération Pêche*, operational between 1977 and 1985. Rice was husked in the rice factory of the *Opération Riz* Mopti operational between 1974 and 1991. A slaughterhouse served the needs of the urban consumers of Mopti. In the nineties a small dairy factory was set up processing milk from contracted, specialised cattle raisers. Apart from the last-named, these processing industries were all state-owned and created in the seventies when optimism reigned with regard to the agrarian potential of the Mopti Region. The idea of processing the agrarian produce in the regional town was clearly based on theoretical views on rural centres mentioned above.

Mopti appeared to be an important service centre: several institutions providing services for the urban and rural populations were established in Mopti since independence. Different types of services can be distinguished. Firstly, the already-mentioned public services related to the position in the urban hierarchy. They were all oriented to the urban and rural population of the region. Secondly, the rural development services (ODRs), government organisations with a certain degree of autonomy, and which were product oriented: the already mentioned *Opération Riz Mopti, Opération Pêche Mopti, Opération Mil Mopti,* and *Opération de Développement de l'Elevage dans la Région de Mopti.* Thirdly, the international development organisations such as PAM, UNICEF, Six-es, UICN, SCF. These organisations had opted for Mopti as a operating base for their work aimed at improving the problematic socio-economic situation in the surrounding region. Finally, there were the private services in the form of trade, processing, storage, transport, banks, religious services and recreational facilities. These private services catered to the urban population or derived their location from the infrastructure which connects Mopti to the surrounding areas.

However, the presence of a service does not guarantee its use in general or for the targeted population groups, as Van Teeffelen (1992) observed in another part of Mali. As for the educational and health services in Mopti, a large share of the users were found to come from the city itself. The surrounding areas made use of these services only to a small degree. The immediate surroundings of Mopti did not take much advantage of their relative proximity. As soon as some means of transportation was needed attendance rates dropped sharply. The regional hospital was under-utilised, more than half of the 150 beds were unoccupied since its opening in 1973 (Zoomers 1984). The costs of medication, hospitalisation and transport were significant factors explaining this under-utilisation. The use of postal, telephone and administrative services also was greatly limited to the urban population.

It is to expect that the services provided by the rural development organisations would cater to the needs of the rural population. But not all the rural people had equal access. In practice there appeared to be a certain preference of these organisations for accessible villages and producers with some development potential. The nature of the development message of these organisations, could be criticised as well: it was geared towards increasing production, without consideration of fluctuating circumstances or sustainability.

The private services were mainly of a commercial nature. Since the liberalisation of the economy in the eighties, this type of services has gained importance. Commercialisation, packing, transport and storage were mainly oriented to customers in Mopti, most of them traders. This does not mean that these services were of no importance for the rural economy. On the contrary, efficient functioning services are very important for the relative competitiveness of the products on the regional and (inter)national markets.

Impacts of Economic Stagnation on Urban-Rural Relations: Trade, Flows of People and Services

During the good years, the production of Mopti's hinterland was largely sufficient to meet the demand of the region's population with even a surplus for export. Intra-regional exchange provided a more varied range of food and a supplement to local deficiencies. Supply from outside the region consisted of industrial products, luxury articles and cereals. Cereals, destined for areas of insufficient production, the northern regions, and areas within the Mopti Region, were imported especially during the period before the next harvest. The long-distance import of cereals was compensated by the long-distance export, especially of livestock and fish, to the south. As such, the exchange of goods on this supra-regional level was more or less balanced.

During the period of drought (from 1983 to 1985, referred to as the second drought) the market functions were directly affected as a result of a smaller production in Mopti's hinterland. The catches of fish decreased sharply. The fish processing industry in Mopti was affected as a consequence. Processing became unrewarding due to the limited supply. This caused the OPM to close the factory. Ever since the fish exported from Mopti is again smoked or dried in the traditional way. Storage activities decreased as well as trade, transport and many other related activities. Millet production had decreased sharply as well. As millet was mainly grown for private consumption, the areas particularly in the dry zone had to be supplied with food from outside the area. During certain years rice production in the polders was almost nil. The yield of rice produced on the flood plain was for a very long period significantly poorer than before. The supply of rice to the modern rice mill in Mopti had never reached the projected level but the drought compounded the situation to the extent that there was inadequate rice to keep the mill in operation. Here as well, the mill was offered for sale. Animal husbandry suffered severely as a result of the drought. Many animals died of starvation and thirst, and many were sold. As a concentration of livestock occured in the Delta - one of the few places where there was still some grass - the supply of livestock at the markets along the Delta was unusually high in the dry years and dropped sharply in the years after. The price dropped and rose accordingly. The response to the drought of the other production activities in the hinterland is not known. The onion crop probably was affected by the drought but craft production might have remained on the same level. The general decline in production in Mopti's hinterland and the low prices of the livestock sold, decreased the purchasing power of the rural population. Nevertheless, the people needed basic foodstuffs in order to survive. An increase in the gathering of edible, wild plants was noted. In the years of poor production, large amounts of cereals were imported from the south. At the same time, the collection of products from Mopti's hinterland decreased sharply and greatly upset the equilibrium in long-distance transport. Trucks had to wait a long time in Mopti for return cargo or they returned empty. This increased the costs of transport considerably.

The imbalance in the flows of goods has had considerable consequences for the areas concerned. For many years, especially in the hinterland of Mopti, cereals were imported from outside without a reciprocal outgoing flow of goods. Under extreme circumstances, free food was distributed by the ODRs and organisations of international development aid but compared to the total flows of goods it involved small quantities. This situation of imbalance in the flows of goods was accompanied by an impoverishment of the rural population. The people sold their valuables (jewellery and livestock) and even their means of production in order to purchase cereals. The environment was exploited by the cutting of wood etc. Part of the population emigrated - temporarily or permanently - in search of paid jobs.

Intra-regional trade was less unbalanced but also suffered: in both directions (from the Delta zone to the eastern part and vice versa) the supply of goods decreased when the production circumstances became unfavourable.

The supply of services was probably the least affected by the economic deterioration. In the years of disaster food distribution was organised by the Social Services Department. In general however the public services did not change much. Even the collection of taxes continued, though they became increasingly difficult to collect. On the other hand, the use of the public services declined as a consequence of the decreased purchasing power of the population. Demand and supply of the public services became even more disparate.

The services of the ODRs were also used to a lesser degree. The people did not see the benefits inherent in the ODRs' objectives of intensification and production growth and the ODRs were too inert to respond to the changed situation. Through the networks of the ODRs, however, part of the rural areas was supplied with food aid and a few 'Food for Work' projects were started to relieve the most serious distress. Only the international development services clearly increased in response to the worsened circumstances in the town and the surrounding areas. The fact that the ODRs were impotent at the time of the second drought has stirred up the debate about the continued existence of certain ODRs. The result was an obvious reduction of the ORM's and OPM's scope of tasks - the economic tasks especially had to be given up - as well as the discontinuance of the OMM.

Mopti's Functions for its Hinterland in Comparison with Rondinelli's Views
Mopti indeed had many of the functions mentioned by Rondinelli (see above), but the performance of some of those functions was moderately. The use of public services was mainly confined to inhabitants of the town. Limited use of health care facilities and low participation in education were noted. This limited use only worsened with the decreasing buying power of the rural population at times of drought.

Processing of rural produce stimulated by government investment took place in a few modern enterprises, which functioned most of the time below their capacities. This modern processing was doomed to come to an end when rural production in the region decreased. The present small-scale enterprises were not oriented to the processing of regional produce.

Among all functions those of collection, storage, transport, and distribution were important and showed a certain dynamic response to changing conditions. Through these functions Mopti was a point of connection between different subregions and between its own region and the outside world. In the period with good production conditions the urban activities in these fields indeed occasionally stimulated market production in the surrounding areas. The opportunity to sell fish at an attractive price in Mopti had led to an intensification in fishing and, temporarily, to an increase in the earnings of the fishermen. The market production of cereals was influenced positively by the presence of wholesalers operating from Mopti. In periods of bad production conditions the activities in the fields of transport, storage and distribution were very useful for the survival of urban and rural populations. It is doubtful, however, whether these functions had led to a lasting development and/or permanent increase in income in the region. Exhaustion of the resource base was evident not only in fishing but also in agriculture in view of the apparently escalating problems of decreasing soil fertility of pastures, rice and millet fields (Diagnostic Régional 1985; Cissé et al. 1990).

Although in many respects Mopti fits Rondinelli's description of a secondary city with its corresponding functions, it cannot be concluded that Mopti played an important developmental role for its hinterland. Under relatively good production conditions it probably did in a limited way, but when agrarian production stagnated the urban economy was of little use and suffered more from the stagnation than it contributed to any improvement.

Summarising, it can be stated that the drought has provoked Mopti's regression to its original functions i.e., a centre of collection, transfer and distribution. From a service centre in the larger sense it again became a 'market centre'. Modern processing has partly been dismantled, the services for the rural population eroded. The services set up to stimulate modernisation and development have not been resumed after the drought. The traditional activities, trade and craft, were also severely affected by the worsened conditions, but they survived. The contacts between the town and the surrounding areas have decreased and, at the same time, the intensity of contacts has been to a lesser extent dictated by the government.

Characteristics and Relations Between Urban and Rural Enterprises
In the discussion of various theoretical views on town-hinterland relations, Schatzberg was quoted as an exponent of the view which ascribes an exploitative role to cities in relation to their rural hinterland. To what extent did Mopti profit from the relations it maintained with its surrounding areas?

As far as the services were concerned, Mopti surely benefited from the employment created and the money spent. The interactions in the collecting and processing fields, however, were a more complicated matter. A surplus transfer from small-scale farmers to capitalist urban traders and manufacturers as suggested by Titus (1991) could be supposed, but in reality the linkages were far more complex.

Agricultural activities in Mopti's hinterland in general were small scale and mainly oriented towards subsistence production; (eventual) surpluses were sold at the market. Among these activities, fishing was the most capitalistic, i.e., a significant share of the fish was brought to the market and relatively much capital was invested in the means of production (boats, nets, engines, diesel) which were frequently bought on credit from the traders in Mopti. Modern rice production in the large-scale and small village polders also was (largely) oriented to the market. Inputs obtained on credit, had to be paid back after the harvest. Traditional rice production on the flood plain on the contrary was mainly for own use; few external inputs were used. In the case of animal husbandry, a more market-oriented production was gradually adopted, as indicated by the widespread use of veterinary products. However, by far not all the livestock was kept for its market value. Milk production around Mopti for the dairy factory, though, could be called capitalist: external inputs were used in the form of fodder and cake besides veterinary products. Millet production was probably the least capitalist form of production and, to the largest extent, geared towards subsistence production. Wage labour was employed in all these forms of production, especially as extra labour during the peak seasons. Commonly, this type of wage labour was provided by people who themselves were producers.

In the period of research, the processing of products from the surrounding areas of Mopti mainly took place in a few (large) state enterprises. In spite of the fact that these enterprises did not aim at making profits, they can be included in the capitalist sector because of their market orientation, the use of wage labour and other purchased means of production.

Was there competition between these large-scale enterprises and small-scale rural or urban enterprises? Modern large-scale processing of rice did not develop before market-oriented production started in the polders. Small-scale processing in the villages always took place but concerned rice for private consumption and the eventual surplus production of traditional rice for sale. The share of small-scale processing probably has increased due to the impact of the liberalisation of the grain market. Millet is not processed, processing is done only at the household level for own consumption. In contrast, fish has to be conserved, otherwise it has to be eaten immediately. Competition in processing between the modern large-scale fish factory and small-scale processing (drying or smoking) by fishermen's households however, did not take place. Fish that was brought in over a large distance had to undergo primary processing in the catch area to make it less perishable. Sometimes it was processed again in the factory in Mopti. Fresh fish brought in from a shorter distance fetched a higher price. A part of this fish was also processed in the factory at Mopti. In the research period demand for processed fish was sufficiently great to prevent competition between different forms of processing. Livestock left the area on the hoof, only a small part was destined for the town's market and arrived in the slaughterhouse of Mopti. In conclusion it can be said that there was no competition between the small scale or non-capitalist and the modern or capitalist manufacturing sector in the Mopti Region.

In the research period, the collecting of products from the primary sector was carried out by traders (fish, livestock, cereals) and by government agencies (cereals). To what extent did surplus appropriation take place and by whom?

In the good years, catching and trading of fish were lucrative activities. The fishermen households were known to be relatively prosperous. The lasting decrease of the catches have put an end to this prosperity. The fishermen did not buy their equipment as frequently as in the past and their creditworthiness decreased. Consequently, the traders would also be affected and previously easily made earnings were lost.

Besides this unfavourable trend, the prices in the fish trade fluctuated sharply. On the one hand, this was caused by the seasonal supply. On the other, this was due to the fact that the fish was exported on a large scale whereas the supply to the domestic markets was in small portions. The prices paid in the fishing areas differed from those in the fish markets. The highest prices were paid in Mopti. As such, the trading of fish in the catch areas which were far away from Mopti involved financial risks for both the producers and the collecting traders.

Livestock could change hands several times before arriving at the user or consumer. Generally, the direction of the livestock trade was southward. Via collecting and regrouping markets, the cattle was taken to livestock markets where part of it was bought up for export. During this process of buying and selling, the prices would increase. They did not only fluctuate fairly sharply in space but also in time. Apart from fluctuating demand due to the infrequent wholesaling at the end-markets, price variation in time was caused by the seasonal supply. The protracted after-effects of the alternating periods of dry and rainy years caused large fluctuations in price, which were especially disadvantageous to the producer-sellers whose livestock was sold at low prices in dry years. They had to pay high prices when the circumstances improved and when they wanted to build up their herds. As the price movement of millet ran contra to that of livestock, such periods of drought meant a considerable loss for the pastoralists. In the trade of livestock, the intermediaries received a fixed commission from the buyer and sometimes a small amount from the seller as well. The profits of the cattle traders appeared to vary greatly. A successful exporter would earn about five to ten percent of the purchasing value, but a less successful exporter could also lose money on a transaction. In itself, the profit margins in trade did not seem extremely high but the livestock exporters especially were found to make large profits due to the large number of animals they sold per convoy at the foreign market. The profits of the traders who sold at the domestic market were not as large.

In the past, the trading of cereals had been controlled by government agencies such as the OPAM, *Opération Mil Mopti* and *Opération Riz Mopti* which paid a fixed rather low producer price. Evasion of obligatory supplies and sale on the black market would bring in more money but were very risky undertakings. Since the nineties, the ODRs - as far as they still exist - have had to hive off their commercial activities. Thus, the farmers are again dealing with the private traders. In the sale and resale of cereals, the traders' margin of profit does not seem to be very large either. It is difficult to gain a clear insight into this trade, especially as other means, such as incorrect assessments of the weight and quality, could increase the margins. The producers did not seem to earn more money since the liberalisation of the cereal trade, contrary to the expectations of some of its advocates.

Trade involving collection and distribution took place on several levels. On the lower level, there were no activities which could be termed capitalist: family labour was by far the most important production factor, very few of the means of production used were purchased from the market and, usually, there were no profits at all. Many traders would even work for very low remuneration. On this level competition was extremely fierce. The most important consideration for involvement in petty trade was the lack of other opportunities for earning money. On the higher levels of trade, making profits was the main objective; purchased means of production were used and wage labourers were employed, although labour performed by apprentices and unpaid relatives was found as well.

It was very difficult for traders to work under strongly fluctuating production conditions and in an ever-changing government policy environment which exerted direct consequences for supply and demand and for price levels. Also the poor infrastructure, bureaucracy and corruption, and the lack of information increased traders' risks. Small traders reacted by investing as little as possible; they bought up small quantities and resold them quickly. To bridge distance and time, usually a very large number of traders were involved. Their risks remained small and so did their profits. They could hardly be regarded as capitalist entrepreneurs. Big traders, especially those involved in long-distance trade (export of fish and livestock), had to make larger investments in storage and transport, and larger purchases. Compared to small traders they made big profits but due to the fact that they needed to invest large sums, they ran relatively great risks. Theoretically, the ratio between profits and risks was favourable for the big traders because of the scarcity of capital in Mali and especially in Mopti. Credits at low official interest rates only were accessible to traders who put up collateral in the form of storehouses, stocks or means of transport. The limited access to a certain level of trade arising from this practice could cause the profits to be comparatively large.

These data lead us to make a slight modification in the Titus model (1986) to fit the conditions of our research area. The most important interaction of the collecting activities in the rural areas took place between the 'more capitalist rural producers' and the trade sector. A very large number of small traders were involved in this trade; some of them traded at their own risk, and some were employed by a big trader. When involved in the collection (and also distribution from the town to the hinterland), small traders were predominantly active in the rural areas whereas the more capitalist traders were located mainly in urban areas (with the exception of some big capitalist livestock traders who operated and resided in the rural areas). The big traders in fish, livestock and cereals certainly made profits, and surplus was transferred in this process of trading from rural to urban areas. Some traders benefited from the fact that the production capacity was affected by the drought. The big livestock traders could make a good deal of money out of the large quantities of livestock offered for sale at relatively low prices which they could sell in other areas whose economic situation was more favourable. Due to the sharply risen local price of cereals, the big cereal traders who imported cereals or organised the transport of emergency relief could also make large profits.

Surplus was also transferred by the below-capacity and/or bureaucratic functioning of the processing industry. The processing capacity was based on an optimistic extrapolation from the favourable production figures in the sixties. Nobody really benefited from this transfer of surplus, except probably the employees who did not have to work as hard for their wages. This uneconomic functioning, exacerbated by the fall in agrarian production during the drought years, resulted in the closure of a number of processing companies in Mopti.

It could be supposed that, in the case of a fairly continuous production growth in the agrarian sector, the larger capitalist enterprises (in the rural areas and in the town) would probably gain the most as they are more capable of releasing capital for the required variable and fixed production costs than the enterprises which operate on a small scale. However, in the case of economic stagnation, this is less obvious. In Mopti and its surrounding areas the boundary between losers and winners did not follow the (sometimes vague) dividing line between (state) capitalist and non-capitalist enterprises. Much depends on the volume of the investments in fixed production assets such as machines and buildings. In order to survive and make a profit, therefore, it is a prerequisite to be able to take advantage of rapidly changing conditions of supply and demand.

Mopti's Production Structure

The preceding sub-sections illustrated that the town of Mopti fulfils many functions. The town is an important administrative and service centre, an attractive place of business for traders and craftsmen, and an important migration destination as well.

The predominance of these functions are typical for many towns of colonial origin in Africa. Initially, urban settlements of this type were administrative, military and trading stations by means of which the colony was annexed to the economy of the mother country and later to the world trade system. These settlements seldom developed into productive industrial centres and their economic base remained weak.

In West Africa the economic centre moved from the interior to the coast as a result of colonial penetration. Consequently, the most important colonial towns emerged in the coastal zone, and many inland (Sahelian) areas and towns lost their functions and stagnated. This situation did not change until after the Second World War and especially after independence around 1960 when urbanisation in the Sahelian countries regained momentum. However, during this period the urbanisation process was hardly or not accompanied by industrialisation. As a result, the nature of production of post-colonial cities and secondary towns in Sahelian countries has not changed structurally. In many of these urban centres, labour is still concentrated in the public sector, trade and personal services and only a very small percentage of the labour force works in the so-called formal sector (World Bank 1979; OECD 1988).

As towns grow, their production structure changes and consequently also the relation between the informal and formal sector, or in a somewhat different perspective between the petty commodity and the capitalist corporate sector. Santos (1975), for example, argued that the less important an urban settlement is, the smaller the relative part of its formal sector. Davila and Satterthwaite (cited in Titus 1991) concluded that in small towns as a result of insufficient demand mainly informal activities are found. Other authors stated that in large(r) urban centres the informal sector continues its existence as it is functional for the formal sector (Hugon *et al* 1977; Coquery-Vidrovitch 1988) and/or both sectors do not directly compete with each other and each serve a separate part of the market (Pederson 1990).

In this subsection, we will look at the production structure of Mopti within the context of the above-mentioned views. What can be expected of the nature of its production structure under the conditions prevailing in Mali in general and more specifically in the Mopti Region?

On the one hand, positive economic developments as mentioned already in the preceding sections of this chapter have occurred and could have promoted growth in the urban economy. On the other hand, factors such as periods of drought and the impact of structural adjustment policies - implemented since the early eighties - could have hampered economic growth. Similarly, the influx of large groups of immigrants from the drought-stricken rural areas could have undermined the subsistence opportunities of small-scale enterprises and the urban self-employed because of the increase in competition.

Given its place in the urban hierarchy and the many functions of the town, Mopti could be expected to have an important modern formal sector. Firstly, as illustrated in sub-section 3.2, many public and private service and development organisations are located in Mopti, has some modern processing industries, and is the operational base of important trade and transport companies with international linkages. Secondly, because of the growth of the town - especially during the sixties and seventies - and, consequently, the extension of the market, local craftsmen and traders from the petty commodity sector could increase their activities, too. A concomitant diversification of the local economy, which accompanies this growth, could be a probable result (Pedersen 1990). At the same time economic growth could have stimulated processes such as specialisation and professionalisation of firms, even leading to the development of enterprises which could be classified as belonging to the formal sector. However, other developments are possible: accumulation of capital does not always have to lead to modernisation of production and/or management or to the development of (large) enterprises with a more formalised status and formal labour relations. Entrepreneurs can also choose to spread their risks and to invest their capital (from profits) in other activities than in their original enterprise (MacEwen Scott 1979). This is more obvious in situations of economic insecurity and - as already indicated - this was the case in Mopti during the seventies and eighties.

Before we examine the nature of Mopti's production structure and its enterprises, we will first look at the employment structure to determine the importance of the various economic (sub)sectors for employment and income. After paying attention to the labour force, we will subsequently, discuss the types of enterprises, their management and their economic linkages.

The Urban Employment Structure
Figure 3.4 shows the division of the economic activities of Mopti's working population according to sex and economic (sub)sector. It confirms the image of Mopti as a trading centre. More than one in three performed activities were related to trade. Craft (17%) and agriculture (16%) were the next most important sources of employment but the total number employed in these sectors was slightly less than the number employed in trade alone. The government sector occupied fourth place, accounting for ten percent of the employment. Relatively few people were employed in construction (4%), personal services (6%) and transport (4%). The tertiary sector was clearly the most significant one in terms of employment of men and particularly women in Mopti.
Further division of the main activities according to type of employment and type of income provides an insight into the importance of these sectors for security of subsistence. The results indicate the minor importance of wage labour. Three in four activities were carried out

in self-employment or as family help. Consequently, most of the people had incomes which were insecure and fluctuating, especially those who worked in the craft, construction, trade and primary sector. How sharply income can fluctuate can be illustrated by the earnings of a maker of banco building bricks. In the rainy season, when the demand is low, his income hovered around 5000 F.CFA a month. In the dry season, however, his earnings could amount to 20,000 F.CFA or more.

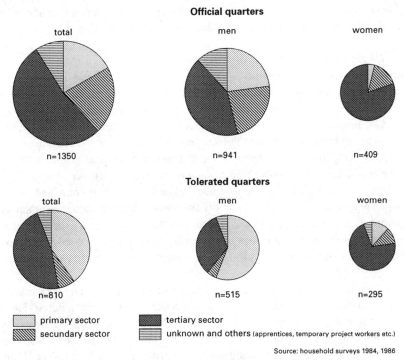

Figure 3.4: Economic activities (main and additional activities) of Mopti's working population according to sex, economic sector and residential area (in %)

In the economic (sub)sectors characterised by a relatively high percentage of wage labour such as government services, transport and personal services sectors we found the highest percentage of jobs with a guaranteed income. Together, however, these sectors accounted for only a small percentage of total employment. Besides, the levels of these more or less fixed earnings varied greatly. High wages were earned by truck drivers; due to several bonuses, their income could amount to 75,000 F.CFA a month. Their apprentices, however, received considerably lower wages - between 10 to 15,000 F.CFA - but these were still higher than those earned by most of the maids, laundry girls and the casual construction workers.

The employment situation of the people in the tolerated quarters was more insecure than that of the other urbanites (see also Krings 1987). These usually recently arrived migrants found it difficult to get a place on the labour market and to find enough work to fill a full working day. They had to create their own work or to try and earn some money as casual workers. Half of the workers must be categorised as part-timers - i.e. people who do not work every day or who work less than five hours a day. The percentage part-timers in the tolerated quarters, exceeded twice that of the other urbanites. The greater subsistence insecurity of the squatters provided the impetus for - in comparison with the other urbanites - a higher labour participation, especially among women, more frequent sideline activities and begging. This unfavourable employment situation of the migrants living in the encampments shows that the capacity of Mopti's economy to absorb new labour was rather restrained.

Enterprises: Dominance of the Petty Commodity Sector
The enterprises in Mopti can be divided into five levels or types, classified hierarchically according to various important indicators: the level and security of income of the people employed, the degree of organisation and the level of investment. The latter two indicators are measured by form and structural characteristics such as registration status, application of technology, kind of business premise, the keeping of accounts, the engagement of wage labour or apprentices, the involvement of the entrepreneur in production or distribution activities, the availability of stocks and the amount of invested capital per worker.

At the bottom of the hierarchy are the 'undertakings' of the self-employed individuals. This type of enterprise has no registered and official status, although sometimes there is some kind of authorisation in the form of a hawking licence for street vendors, or a seller's ticket bought daily by market traders. In general the self-employed have no special work-shop, they rely on their own labour, make use of simple techniques and tools and have little capital available.

Household units are the second type of enterprise. In this type of enterprise two or more members of a household perform activities which relate to each other, but which do not necessarily belong to the same economic sub(sector). The most important difference between this type of enterprise and that of a self-employed is - besides the probably more complex degree of labour organisation - the ability of particularly household units in adapting to changing production conditions by means of altering their production behaviour as well as their reproduction and consumption decisions (Schmitz 1982, pp. 431-432). With respect to structure and form characteristics there are no other great differences between enterprises run by a self-employed worker and that of a household unit.

Owner-operated family firms are enterprises in which the owner, whether together with family members or not, actively takes part in the production or distribution process, assisted by apprentices and/or paid wage labourers. In any case, an owner-operated firm differs from a household unit in the extra labour engaged from outside the family and in the separation of living and working space. On the average, slightly larger capital investments have been made (without any credit from the bank) and increasingly more of these enterprises acquire official status as they pay the business licences and taxes. Generally, no accounts are kept. The degree of organisation is higher than that of the two already described types and entry to this level is less easy because more investments are needed.

Private enterprises in which the owner and his household do not provide the actual labour on the work floor but instead only supervise and/or manage the work are considered owner-managed enterprises. This type of enterprise includes a range of firms with owners who have invested their capital in the means of production and who employ people on a permanent basis to do the work. Occasionally casual labourers are also employed. The amount of capital invested in these companies can vary considerably; bank credits may be used to engage in this form of production and organisation but are not that common. Also the size of the companies can vary considerably: this category includes very small enterprises (with one or two employees) as well as very large ones.

The last type at the top of the hierarchy is made up of manager-operated companies run by professional managers. In principle, it concerns both state and private companies where (almost) all the employees are employed on a permanent basis. In Mopti, this type includes the government institutions and non-profit institutions.

Of all types of enterprises, the individual and household enterprises clearly display the most number of characteristics of the petty commodity sector, such as low degree of organisation, low capital investments, unauthorised status. Enterprises run by managers are the most 'corporate' in nature. The first three types of enterprises fall under the category of informal or 'petty commodity' sector. Companies and institutions run by the professional managers belong to the formal or corporate sector. Enterprises of the owner-managed category, however, can belong to either the former or the latter sector.

Looking at the enterprises in Mopti, an extremely pyramidal structure can be observed: there were many small-scale 'undertakings' of self-employed, and only a few large-scale organisations - private companies and public institutions. This means that in Mopti, the petty commodity sector is by far the most important sector for employment.

This pyramidal structure was a result of the economic environment which can be characterised by weak purchasing power, fluctuating demand and a scarcity of capital. In such an economic climate it is necessary for entrepreneurs to keep their overheads as low as possible, to abstain from making large capital investments, to ensure that capital is well managed and safely utilized and to build up personal networks with suppliers, consumers and trusted agents. In short, they have to limit the risks of management.

The necessity to keep the overheads as low as possible stemmed, primarily, from the absence of regular wage labour in the economic sectors which were most important in terms of employment. The majority of the enterprises consisted of the self-employed: thousands of market traders, street vendors, porters, brick makers, laundry women worked for themselves. Even in the more formalised part of the craft and trade sectors viz. craft workshops, shops and wholesale stores - a quarter of the enterprises was one-man business operations. The number of firms with more than ten workers was very limited in Mopti. The somewhat larger firms were found in freight transport, where eleven percent of the firms employed more than ten employees. In contrast to workers in trade and crafts, all the employees in freight transport and those involved in the transport of passengers were paid for their labour. Craft and trade, on the other hand, still operated the apprentice system with a comparatively long apprenticeship and with considerable use of unpaid family labour.

A second way to keep overheads low was to do without special workrooms or by sharing one. In all economic sectors, we found indications of costs being minimised in the accomodation of the business activities. Among the companies in the trade sector, including the activities of the self-employed, the majority operated at the markets sites, at home and alongside the streets under small reed shelters. Craftsmen generally worked in workshops although these workshops were not owned by the user but rented in most cases. Often several craftsmen and their apprentices could be at work in the same workshop without any form of organized cooperation, even if they did the same kind of work.

Finally, overheads can be kept down by minimal investment in tools and machinery and by putting off their replacement for as long as possible. Tools (e.g., sewing machines) were even rented, occasionally for the duration of the peak season alone. Of course, the necessary investments vary greatly per activity. A repairman with a welding apparatus or the owner of a pick-up had considerably more expensive tools than a leather worker or a weaver or a porter who rented a push-cart. The fact that the replacement of tools or equipment was delayed was most evident in the transport sector, where, on average, the fleet was old (more than half of the trucks were 13 years or older) and often off the road for repairs.

The aim to limit the risks as much as possible, also became evident in another aspect of management, which at the same time pointed to capital scarcity. This tendency showed in the limited trade stocks of the market traders, vendors and craftsmen and in the tendency of shopkeepers to replenish their stock of certain products only when the old stock has been sold out completely. Therefore, there were occasions when certain articles were not available in the city for a couple of days.

The major proportion of the workers in the craft sector did not have raw materials or final products in stock and worked only on order from their customers. In many cases, the customer made a down payment which enabled the craftsman to purchase base materials or the customer supplied the material himself. From the fact that the base materials used by the craftsmen were not owned by them, it can be deduced that these self-employed did not have full control over all of their means of production.

The aim to keep down the risks as low as possible also affects the organisation of some of the enterprises classified as owner-operated and owner-managed. In these businesses, with comparatively more invested capital, risks were spread as well. The key characteristic here was a specific type of labour division. This division does not entail specialisation within the companies, rather it implies a division of tasks: the (relatively) independently operating workers are entrusted with a (particular) part of the organisation and/or the implementation of the production or service process. In the centre of the network is the entrepreneur of substance - e.g. a wholesaler, haulier or investor. He is the driving force behind the activities of others, because of his comparatively strong financial position. Such labour division was found in most sectors of the economy: trade, transport, construction and even sometimes in craft.

Furthermore, this type of labour division was not restricted to owner-operated or owner-managed firms alone, although it was most often associated with these types of enterprises. Even in the cases of companies managed by professional staff and undertakings of self-employed individuals such a type of labour division was occasionally practised. Traditional weavers, for example, working alone contracted the manufacturing of warp to others. In the case of the slaughterhouse, forty independent butchers and their helpers worked side by side with some twenty permanent employees. The work of the butchers consisted of chopping up the carcasses of their - bought on the market - cows which were first slaughtered by the permanent workers of the slaughterhouse.

The above analysis makes clear that there are very few enterprises which are in full control over the whole range of activities in any production and distribution process. It is obvious, therefore, that there are linkages between the various activities. We will now examine the nature of these linkages.

External Production Linkages

In the literature, there are several views on the nature of the external production linkages of the petty commodity sector. These views differ according to the scale of the subject of the analysis. In studies which analyse sectors or modes of production, emphasis is often put on the dominance of the capitalist corporate sector and the dependence of the petty commodity sector. The dependent position of the petty commodity or informal sector is manifested in the comparatively high prices of raw materials purchased from the capitalist corporate sector, and in the low prices of products, services or labour supplied to this sector. This unequal exchange results in the creaming off of surpluses and the accumulation of capital by the capitalist corporate sector. Besides, the corporate sector shifts the risks of a fluctuating demand on to the petty commodity sector via mechanisms like sub-contracting and out-work systems (Coquery-Vidrovitch 1988; Hugon et al. 1977).

In studies where labour or the micro- and small-scale enterprise is the subject, the focus is on another kind of dependence. In this case, dependence implies that many of the so-called self-employed are not completely autonomous: they do not fully own their means of production; nor are they entirely free to choose their suppliers and/or customers. With respect to the self-employed, Bromley (1988, p.167) distinguishes two kinds of self-employment: true self-employment and dependent work. The so-called dependent workers are those who are dependent on others for their means of production and/or raw materials. These dependent workers, however, are not by definition always dependent on companies in the corporate sector but they can just as well be dependent on wealthy entrepreneurs in the petty commodity sector, such as traders.

The third type of dependence concerns the network dependency and refers to the necessity to establish and maintain networks to guarantee the functioning of the enterprise and to minimise risks (Roberts 1978; Lowder 1986). By building up personal networks which are maintained both serve to provide examples by strategies such as mutual indebtedness, credit loans and assistance, an entrepreneur, self-employed or not, secures the supply of raw materials or merchandise and/or the purchase of end products and services. These networks are based on mutual trust and mutual dependence. It is not surprising that frequently these networks include family members and people from the same ethnic group or area of origin. Often the goods or services acquired through the network are not the cheapest possible. On the other hand, the entrepreneur will be given credit or assistance at this familiar address when the need arises. This explains why networks constitute a certain guarantee of help in hard times. The more familiy ties are closer or the more the business relationship with strangers are long-standing, the higher the costs of breaking off from such networks. Because of these costs - called shifting costs by Siamwalla (1976) network relations are not easy to end. This creates a situation of quasi-dependent external production linkages.

In Mopti, the majority of the external production linkages was characterised by the latter type of dependence. The second type, i.e. dependence on other entrepreneurs, was also found especially among the traders in the petty commodity as well as the corporate sector. There was hardly any direct dependence of the informal sector on the formal corporate sector.

The small formal sector in Mopti did not practise any sub-contracting or out-work, as was found elsewhere (Gerry 1979; Joshi et al. 1976). This is not surprising given the nature of this sector in Mopti which primarily comprises public institutions and services and, to a lesser extent, agricultural processing industries. This sector saved on labour costs by employing casual workers or temporary employees. The OPAM, for example, paid the porters for loading and unloading the trucks by the sack, and the rice mill employed a maximum of forty-two seasonal workers during the harvest season in good production years.

The industrial firms in Mopti were a kind of *corpus alienum* with hardly any local forward linkages. The machines came from abroad and their own employees have been trained to maintain them, as a result of which there are no linkages with the local repair sector. There were linkages, however, with local suppliers and traders. With the exception of flour for the bakery, raw materials were supplied by producers from the region whereas the customers were partly local and partly non-local users or traders. The slaughterhouse and the bread factory, for instance, had linkages with the local petty commodity sector; the slaughterhouse via the already-mentioned butchers and the bakery with the sales outlets in and outside the town.

Only through its use of industrial base materials and tools did the craft sector in Mopti have linkages with the corporate capitalist sector elsewhere in Mali or abroad. In Mopti, there were only a few crafts which chiefly made use of self-made tools and/or processed mainly local raw materials. The supply of industrial tools and base materials to the craft sector was indirect; via the traders in Mopti's trade district. Nine out of every ten craftsmen we interviewed obtained their raw materials from these traders.

The indirect character of the linkages between the craft sector in Mopti and the industrial sector elsewhere was also expressed in the negligible percentage of craftsmen who had acquired their professional training in the industry. In Mopti, a trade was usually learnt from a master or skills were passed on through the family.

In many respects, trade in and around the trade quarter, and thus also trade around the harbour and in the *Bas-fond* (see figure 3.5), was Mopti's economic pivot. Not only did it fulfil supplying functions for the craft sector but also retail, market and building trade elsewhere in the town depended mainly on it for the supply of merchandise and building materials, respectively. In addition, trade in the trade quarter near the harbour was linked with the craft sector because the traders purchased the end products of both the truly self-employed and the dependent artisans.

In Mopti and the city surroundings, we found several types of activities geared to the production of commodities for trade by intermediary traders. Products such as charcoal, processed fish, blankets, clothing, earthen pots, reed mats and baskets, sandals and water bags for wells were bought up by the specialised traders for sale in Mopti or traded to other parts of Mali. The producers of these goods usually had a regular buyer although they did not share a direct dependency relationship with the intermediary trader.

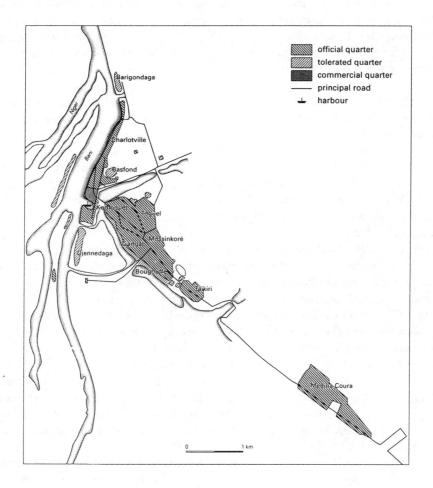

Figure 3.5: Location of main types of quarters in Mopti

Dependent workers were also to be found, e.g. among the tailors and especially the weavers. A large majority of the numerous weavers in Mopti were seasonal migrants from the surrounding rural areas who only spend a part of the year in Mopti. It was advantageous to these traditional craftsmen, to have a regular business relationship with a (woman) trader who supplied them with yarn, paid them by the piece, and occasionally also arranged board and lodging. In this way, the weaver did not waste time in finding a place to sleep, obtaining raw materials or selling his products, and he could do without a starting capital. Similar to the many craftsmen who are self-employed but with regular sales addresses, the dependent weavers were also free to choose other customers. However, for them too the shifting costs were high.

The major part of the linkages between traders and craftsmen consisted of personal networks on a voluntary basis. As long as both parties continued to see advantages in such voluntary relations, these continued to exist. The trader-customer was ensured of a constant supply from her/his regular producers. In return, he or she was prepared to give the producer an advance on the next delivery when the latter needed capital urgently. Similarly, the trader-supplier of materials or tools could supply the producers with these goods on credit and consequently secure a regular source of sales. If the credit given resulted in a debt that could not be repaid, the tie between the supplier and customer soured and the relation lost its voluntary character. This can be illustrated by the relationship between fishermen and suppliers of nets and outboard motors. Fishermen, who were incapable of repaying their debts, were forced to sell the fish via their creditor. Generally, the latter paid the fishermen far too little for the fish they supplied as a result of which, consequently, the fishermen had to pay 25-30 percent more for their equipment than they would have otherwise (Diop 1971, p.111). If a fisherman failed to pay off his debt, he could, at worst, even lose control over his means of production.

Capital is scarce in Mali and credits, as a result, are not easily given, especially not long-term loans. In the linkages between enterprises, credit plays a specific role. It is common practice to supply only a small part of the goods on credit in the case of large deliveries and to give goods or raw materials on credit for only a very short period in the case of small quantities. Contrary to our expectations, entrepreneurs at the top of the enterprise hierarchy gave credits less often than entrepreneurs lower down in the hierarchy.

The fact that larger traders, for instance, were reluctant to sell on credit indicates two things. In the first place, it indicates capital scarcity among these entrepreneurs who, after all, have invested a sizeable portion of their money in their trade stock. In the second place, it indicates sufficient sales potential as a result of which a trader does not feel that he needs to secure his market share. This situation occurs particularly in monopolistic or oligopolistic markets. In Mopti both reasons apply; depending on the kind of goods and the extent of competition.

In Mopti there existed a gap in many respects between large owner-managed firms and the rest of the enterprises. A limited number of large firms in Mopti were, without any exception, owned by well-to-do families with interests in several economic activities. These activities are closely linked to each other as a result of which firms like these largely controlled the production and distribution processes. All the other smaller enterprises, ranging from the self-employed to the small owner-operated ones, were largely dependent on others as far as their external production linkages were concerned. This dependence did not so much concern the servitude of the informal with regard to the formal sector, as described in the literature, but rather the relations between producers and traders and among traders. This dominant position of the traders in these networks emphasises again the importance of trade to Mopti's economy.

Effects of Economic Growth and Stagnation
In the first three decades after the Second World War, Mopti's economy developed

favourably. Money could be made in trade and transport, and other sectors such as craft and construction benefited from the increased sales potential and orders as well. On the one hand, this positive development resulted in an enormous growth in the number of small-scale enterprises. Roughly, the number of licences for enterprises with the smallest turnovers tripled between the late fifties and the eighties whereas the number of enterprises with large turnovers increased only slowly. The growth of the non-registered sector was even larger. On the other hand, capital accumulation had been possible in certain sub-sectors of the economy. Was the money invested in the firm and did it lead to modernisation and increased professionalism in management, even to specialisation? Or were the earnings invested in other economic activities as a result of the strategy geared to spreading risks, thus leading to diversification?

According to our findings, in times of economic prosperity, the entrepreneurs in the economic sub-sectors where demand was high dared to invest in their own company thus increasing their business opportunities. In all sub-sectors of the economy, successful entrepreneurs were to be found who were willing to invest. This conduced to changes in the performance, and expansion of the business activities. The successful trader or craftsmen invested in shop premises or was willing to pay relatively high rents for demanded locations: he moved his activities to a (work)shop for instance along the main roads, near market places or in the trade quarter near the harbour.

Trade and transport are the mainstay of Mopti's economy. Therefore, the most capital accumulation was found in these economic sub-sectors. There was also growth in certain craft activities, but in general, the capital accumulation in Mopti's craft had not been large. Only fifteen percent of the workrooms investigated contained tools worth over half a million F.CFA in investments. Petty commodity producers could be divided into those engaged in old and new crafts. Capital investments were particularly made in transformed old (a tailor now working with a sewing machine) and in new crafts. However, there were large differences in the investment between enterprises involved in a similar type of craft. One car repair firm, for instance, could acquire electrically driven tools worth hundred thousand F.CFA, whereas in another, only the ordinary manual tradesman's tools were used. A traditional smith employed only simple tools but a specialised sheet metal worker had invested in considerably more expensive welding equipment and a generator.

In the cases of further capital accumulation, or of less positive prospects within the particular sub-sector of the economy, the entrepreneurs tended to invest in economic activities which had hardly any relation to their own enterprise; they did this to spread the risks. The kind of investment was often more dependent on the opportunities that arose rather than based on a well-considered rational choice seen in relation to their own business. Looking at the sideline activities of the traders with shops and storage facilities and masters of craft workshops, we found that half of these people also invested in other economic activities. Investments in the primary sector were dominant but money was also invested in capital goods which were subsequently hired out to others. Money invested in this way was used for, among others, sewing machines, cereal mills and the construction of houses. Such investments can be interpreted as investment strategies employed to spread risks. This phenomenon resulted in a specific type of 'owner-managed enterprise'. In the case of a number of small businesses the description 'owner-managed' did not really apply. Owners of, for example, a means of transport, sewing machine or cereal mill were not directly involved with management which they left entirely to a manager, tenant or intermediary. These owners were more concerned with investment for the best possible returns.

When the business efforts led to a considerable accumulation of capital, it was followed by the wish to increase control of the means of production and to invest in economic activities linked to the original firm. This last step, however, was limited to a small number of enterprises. The possibilities for investment were so unequally distributed that it led to the earlier-mentioned hierarchy of enterprises. In fact, enterprises of substantial size developed only in the trade and transport sectors.

During the years of prosperity in the fifties and sixties it had been possible to make large profits, especially in fishing and the fish trade. Especially in these branches capital accumulation led to professionalism and later to specialisation. Earnings were invested into the same sub-sector of the economy: fishermen, for example, invested in equipment such as outboard motors and imported fishing nets. In time, investments were also made in additional or bigger means of transport. Among the *pirogue* owners, there evolved a distinction between those who worked as fishermen and those who used their boats for transport. Those in the latter group gradually specialised in the transporting of passengers or goods.

In road transport, we saw a similar development in which there was not only a differentiation in the transport of passengers and freight but also a specialisation according to the type of vehicle and transportation distance. Firms with pick-ups or *camionettes* concentrated particularly on intra-regional traffic; companies with trucks were involved in interregional and intra-regional transport over metalled and unmetalled roads; whereas companies with articulated trucks focused on long-distance transport across comparatively good motorable main roads. The latter category of large-scale transport over long distances has especially showed a substantial growth after 1970. The total load capacity of road transport companies from Mopti quadrupled between 1976 and 1986; the increase in numbers of articulated trucks accounted for 70 percent of this growth. The enormous growth in the load capacity, stimulated by the expanding demand for transport as a result of factors such as population growth and the increase of donor aid in kind (food aid), led to a large over-capacity of the trucks judging by the long waiting times faced by transport companies and the decreasing volume of cargo transported per trip.

The development of the transport sector illustrates that, in the case of large capital accumulation, investments were made in the parent company, namely in the expansion of the fleet. In the trade sector, these tendencies were displayed as well, namely through increases in the stock, rent or construction of storage space. Likewise, investments were made in other sectors which were related to the original company. Mopti's most successful companies appeared to be those which combined wholesale with transport. The original company could have been a trade company as well as a transport company. The most striking example in Mopti, was one of Mali's largest importers of fabrics. This family enterprise had, in addition to the textile trade, interests in one of the Malian textile factories, owned fifteen articulated trucks and two tank trucks, ran a petrol station and besides, also had earnings from letting real estate.

Companies with interests in both trade and transport seemed more flexible in their ability to accommodate the changes in the market than those which specialised in only one of the two activities. Their means of transport were not only deployed to transport products in the conduct of their own business but were also rented out to third parties. Besides, the possession of its own means of transport enabled the company to seize attractive business opportunities in the trading of goods other than the original company products. These companies had sufficient capital at their disposal and were creditworthy enough for bank loans, to be able to undertake such ventures. This type of companies also sufficiently invested in maintenance and replacement of their means of production, consequently their fleet was permanently available. This contrasted with others for example, entrepreneurs of small transport firms who owned, on the average, older fleets and who, due to frequent breakdowns of their vehicles resulting from overdue maintenance, were not able to take on as many orders.

In general Mopti's production structure has retained traditional characteristics. Most of the businesses in Mopti operate within very narrow margins, that is to say, their turnovers are low and insecure. However, a process of change has been initiated by modernisation and opening up. In spite of these changes, no structural transformation of Mopti's secondary sector in terms of a proportional decrease of household and craft activities and a growth of workshops and factories occurred. In the tertiary sector, however a transformation could be observed from periodic to permanent trade, and from market trade to trade in permanent business premises as well as a growth in wholesale trade.

Trade and transport are in particular well represented in both the bottom and the top levels of the hierarchy of enterprises. Thus, comparatively, these economic sub-sectors have produced a small number of large basic enterprises, and were also flight sectors for thousands of self-employed people. This is in contrary to the secondary sector in which there are hardly any private formal business ventures and which acts as a flight sector to a very limited extent.

The large owner-managed businesses, which are generally combinations of transport and trade companies, were geared towards profit-making. Due to their larger control over the production and/or distribution processes and the larger availability of capital, these companies were less vulnerable and were not only able to maintain but even strengthen their position in poor economic times. Due to their dependence on the government and donors, the manager-operated companies, which were mainly government and non-profit institutions until the late eighties, appeared to offer their employees less security of subsistence than was formerly thought.

Economic activities have to contend with the low purchasing power of the population and sharp fluctuations in the demand for goods and services. The self-employed and small-scale enterprises could ensure their survival in such an economic climate by minimising overheads and limiting risks. Large corporate enterprises, however, could hardly survive as was illustrated by the bankruptcy of the unprofitable rice mill and the fish factory. In addition to factors like low purchasing power and unstable demand, other factors mentioned by, among others MacEwan Scott (1979), were responsible for the fact that only very few companies in Mopti could develop into large companies: over-supply of micro-scale and small-scale enterprises as a result of which the earnings of the entrepreneurs remained low; scarcity of capital, which accounted for the low investment levels in areas such as new technology; and the tendency to spread accumulated capital in order to spread the risks.

Conclusions

Mopti's development into the most important urban settlement of Northeast Mali has been the result of its favourable geographical situation and of external interventions. Government investment in the transport network enabled Mopti to develop into a (supra)regional trade centre. The town fulfils functions of collection, storage, transit and distribution for its hinterland.

Attempts of the government to assign Mopti more functions and to make it play a more diversified role for its rural hinterland however, met with great difficulties. Investment of the government in administration, education, health care, development services and agriculturally based industry has diversified the economic base of the town but did not drastically change the relationships with the rural environment. The rural areas did not really take advantage of the new services and facilities because of the low rural purchasing power and the poor accessibility and the deficiency of the services provided. In general the urban population took more advantage of the new services.

Fluctuating production conditions and recurrent droughts took a heavy toll on the economic situation in the rural areas. During the drought periods rural production diminished drastically. Some large corporate enterprises especially those manufacturing rural products became bankrupt like the rice mill and the fish factory.

As a consequence of low production levels the buying capacity in the rural areas also diminished. Demand for services and products were limited to the most essential: food. As a consequence urban-rural relations were affected in various ways. The collecting trade and related activities in town dwindled. In contrast, the distribution of food to the rural areas of the Mopti Region as well as to the northern regions became vital.

Mopti played a role which enabled several population groups to survive the drought periods in the seventies and eighties by offering some additional work and income to the impoverished rural population, and by the distribution trade of food via the rural markets. These activities maintained a certain economic base in town. However the urban economy also suffered a lot from the droughts. Demand for goods and services diminished and many enterprises hardly survived. The many self-employed and small-scale enterprises could ensure their survival by minimising their overheads and limiting risks. In general the manager-operated organisations, mainly government and non-profit institutions, survived but they offered their employees less security than was formerly thought. The rural development organisations had to reduce their tasks, and some even ceased functioning.

The oldest economic sub-sector in town, trade and transport, had borne the difficult years in the best manner. Some large companies, generally combining transport and trade activities, even succeeded in making large profits during these periods by investing their capital in better means of transport necessary for the transport of large amounts of food to the rural areas of the three northern regions. At the same time this sub-sector acted as a flight sector for thousands of self-employed people.

Apart from fluctuating production conditions which stimulated the tendency among entrepreneurs to spread accumulated capital in order to spread risks, other factors were responsible for the predominance of the small-scale enterprises in the urban economy and for the fact that only very few enterprises in Mopti could develop into large companies: oversupply of micro-scale and small-scale enterprises as a result of which the earnings remained low, and scarcity of capital which accounted for the low investment levels in areas such as new technology.

Although a favourable geographical situation, demographic growth and a positive government investment policy made Mopti into a commercial centre of more than regional importance, at the same time it has not been able to develop into a city diffusing modern development to its own enterprises and those of the rural environment. Although some enterprises, especially those in the transport and trade sub-sectors, succeeded in modernising their means of production and extending their intervention zone, most production and service enterprises remained very small just enabling their proprietors to survive.

This outcome is not unique for Mopti: many relatively large secondary towns in the Sahel, and also elsewhere in Africa, show the same characteristics and the same potential. Transport and trade are their basic economic activities, and in this respect they fulfil a role for the rural hinterland. Large fluctuations in production and trade over years - whether brought on by physical, political or military events - are unfavourable circumstances for large companies like manufacturing industries or marketing boards with relatively sizeable standing capital. Only flexible enterprises, very small ones on the one hand or rather large ones on the other which can shift capital from one activity to another have the capacity to survive, and in the case of the latter, to grow. Stagnating manufacturing and on-going polarisation in the transport and trade sectors seem characteristic for these towns.

Notes

1. When Mopti or the town of Mopti is mentioned it stands for Mopti and Sévaré.
2. The research in Mopti and its surrounding areas was conducted during the period 1983-1991. The research material consists of data collected by field studies and secondary and statistical data from sources in Mali, France and The Netherlands. During the field studies primary data were collected on households and enterprises in Mopti and its hinterland. Interviews were held with urban and rural households, traders and shopkeepers, entrepreneurs in the field of handicraft, transport and construction, public servants and government officials. These were structured interviews, conversational and group interviews and in-depth interviews with key persons in various fields. In the rural research villages and markets were the most important research locations.
3. Rondinelli listed about twelve functions. In this chapter; though, we will only discuss the economic ones related to urban-rural interaction.
4. Mopti has been an important centre for immigration. During the droughts many refugees from the northern regions and from its hinterland settled in 'tolerated' quarters and in encampments around the city. In this chapter we will not discuss demographic aspects of the urban-rural relationship, notwithstanding their importance, because of limited space.

CHAPTER FOUR
Small Urban Centres in Central Mali.
A Study of the Role of Service Centres in Rural Development[1]

Pieter van Teeffelen

Small towns, urban centres located at the base of the urban hierarchy, often are considered to play a central role in regional development as suppliers of services which facilitate the development of the centre's hinterland (cf. Mosher 1969; Van Dusseldorp 1971; Harts-Broekhuis & De Jong 1985). The research on which this contribution is based started as an inquiry into the linkages between small towns and their hinterland in Mali. In this context attention was paid to the supply and demand of public services (Van Teeffelen 1981). Soon it became clear that, although a wide range of services was offered in a range of small urban centres, the rural population (the majority of whom are farmers) hardly made use of these services. This applied also to rather basic services such as elementary schools and dispensaries. Based on an inventory and analysis of supply and use of public services an attempt was made to obtain an insight into the underlying processes. This essay does not provide a full report of the research, which has been, baptized PROUST[2], rather it is meant as an overview of the main results and a description of the underlying processes.

After a short introduction of the research area in its national context and a synopsis of the relevant public services, the supply and use of public services will be discussed. Then the results of this research will be placed in the wider context of the functioning of small urban centres in other regional contexts in developing countries.

The Regional Context

Mali, one of the Sahelian countries, is among the poorest countries in the world. The country is a republic with approximately 8.5 million inhabitants occupying an area of 1,240,000 sq. kilometres. The northern half of Mali consists of desert where only some Tuareg tribes can survive. The southern "bank" or *sahel* (the Arab word for bank) of the desert is characterised by arid grassland where Fulani graze their cattle, whereas in the more southern parts a savannah landscape can be found. In this more fertile part of the country Bambara form the largest ethnic group. The dry climate, poor soils, lack of natural resources and defective infrastructure constitute such barriers that – even on a longer term – an essential improvement of the welfare and well-being of the population is unlikely.

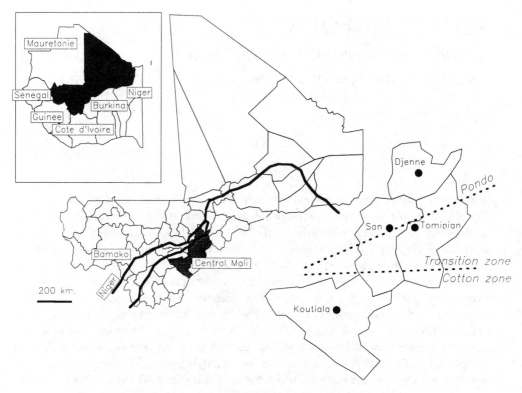

Figure 4.1: Central Mali in its national context

In a demographic sense, Mali is a classic example of a developing country: the thinly populated republic is characterised by structurally high birth rates linked to decreasing death rates. The educational level of the population is low and analphabetism is rather the rule than exception. Public health is in a permanent state of crisis and many households fear the spectre of famine every year at the end of the dry season. The authorities lack both the means and the money to improve these circumstances, let alone abolish them.

Agro-ecological diversity
Although the agro-ecological potential of Mali as a whole can be considered as distinctly low, the country possesses areas with a markedly higher potential. In Central Mali this differentiation can be observed in the form of three distinct ecological zones (figure 4.1).

The *Pondo*, the most northern part of the research area, located in the inner Niger Delta (an area inundated every year by the Niger River and its attributary the Bani) can be considered to posses a high agro-ecological potential. Despite impressive investments in water management and polder development, the production of rice, maize and millet remains remarkably restricted. The fishing industry and cattle raising are perhaps even worse off.

The *Cotton zone*, located in the south of the Central Mali is also blessed with a high agro-ecological potential, the outcome of a combination of deep fertile soils and sufficient precipitation. In contrast to the Pondo, here the high potential leads to a relatively well-developed agricultural production of cotton. Almost every farmer in this area is engaged in the commercial production of these fibres.

The zone in between the Pondo and the Cotton zone –the *Transition zone*- beyond all doubt is the area with the lowest agro-ecological potential. Not only is water extremely scarce (the rainfall is restricted and rivers are absent) also the thin sandy soils are very poor. Except for a limited production of millet and groundnuts for own consumption hardly any other product is grown.

Despite these considerable differences in agro-ecological potential, from a socio-economic point of view, the population seems surprisingly little differentiated in both the interregional and intra-regional sense. The standard of living of all households whether they live in the "poor" Transition zone, the Pondo or the "rich" Cotton zone, hardly differs.
It is therefore difficult to differentiate the "haves" and "have-nots" for the vast majority of households should be considered "have-nots". If differences in living standards occur, it is between the uneducated rural population consisting of farmers and the educated small urban population of government servants and traders.

National development policy
In the three decades since the political independence of the country the national government has not succeeded in improving the often harsh living conditions of the population. Although the improvement of agriculture has been (and still is) one of the major targets of the national policy, agriculture still is largely traditional in nature and commercialisation of agricultural production is –with the exception of cotton production– negligible.

One might however wonder why even in areas with a demonstrably favourable agro-ecological potential (such as the Pondo and the Cotton zone), areas where millions of dollars have been invested in socio-economic services such as health care, education and agricultural extension, hardly any development results have been achieved.

Administrative division
The agro-ecological division discussed above does not fully coincide with the nation's administrative division as can be seen in figure 4.1. The administrative division of the republic reflects its French colonial past. Mali is divided into seven regions (*régions*). The capital Bamako constitutes a separate administrative region. Each region on its turn is divided into districts (or *cercles* in French), 46 altogether. Every district consists of a number of sub-districts (*arrondissements*), totalling 281 at the national level. Every region, every district and sub-district has its *capital*, a town or village where administrative services are concentrated (see figure 4.2 for the districts and sub-districts and the urban centres and villages in the research area).

The Djenné District (part of the Mopti Region) and the smaller northern parts of the Tominian and San Districts (both in the Segou Region) are considered parts of the Pondo. The Koutiala District (Sikasso Region) and the southern edges of the San and Tominian Districts belong to the Cotton zone, whereas the larger part of the Tominian District and the central part of the San District are considered part of the Transition zone.

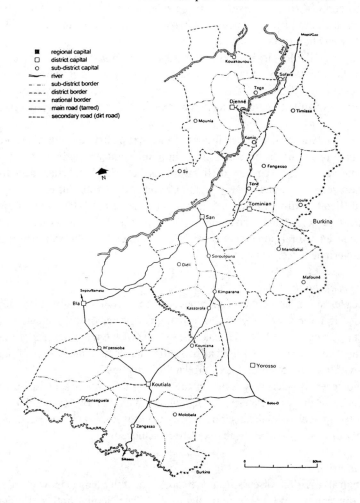

Figure 4.2: Mali, administrative centres and districts in the research area

Settlement structure
The research area has a strong rural character: only 73,500 of the approximately 663,500 inhabitants (roughly 11 percent) live in urban settlements. And even this is to be considered a fairly broad classification of the urban population for in Mali every centre with over 5000 inhabitants is already considered urban while irrespective of size, all district capitals are defined urban. Despite these low urban "thresholds", only four urban centres are to be found in the research area. San and Koutiala have reasonably large urban populations (see also table

1). Djenné at least exceeds the 5000 mark and Tominian obtains its urban status solely from its administrative function. Apart from these "urban" centres only 13 other population centres have 2500 inhabitants or more. More than three-quarters of the total population live in what can be called small villages.

Public Service Supply

In Central Mali, as in the country as a whole, a large range of public services is provided with activities varying from literacy courses to higher agricultural education and from prenatal care to village co-operation activities. This wide variety of services can be divided in five groups: administrative services, two groups of economic services and two groups of social services (medical and educational).

Table 4.1: The settlement structure of Central Mali

Inhabitants	Number of Centres	Administrative Centres
> 5,000	3	Koutiala (35,000 inh.)
		San (23,000 inh.)
		Djenné (13,000 inh.)
2,500-4,999	13	Tominian (2,500 inh.)
1,000-2,499	103	12 sub-district capitals
500-999	218	12 sub-district capitals
< 500	787	1 sub-district capital
total	1,124	

Source: Rép. du Mali, 1988

Administrative services
This group of services is very diverse. The registry office, tax services and public administration are obvious constituents of this group. Also the postal services and a water and forestry agency (in charge of environmental protection) come under the administrative services.
Furthermore, there are several services that in a strict sense do not belong to this group but in practice are closely related to them: the national political party the UDPM (the *Union Démocratique du Peuple Malien*, until 1991 the only political party in the country), and the co-operative movement including the village planning committees "supported" by the government.[3]

Economic services
Two groups of economic services are distinguished. The first group consists of sectoral state

organisations. They have their roots in the socialist past of the republic and were established to organise a specific part of the Malian economy. The SOMIEX *(Societé Maliènne d'Importation et d'Exportation)* is one of these organisations. The SOMIEX was created to manage a national chain of shops for basic consumer goods. The OPAM *(Office des Produits Agricoles du Mali)* was established to control and organise the marketing of the nation's agricultural production. The SCAER *(Societé de Credit Agricole et 'Equipement Rural)* was given the task to distribute agricultural capital inputs (ranging from machinery to fertiliser) and to establish a national agricultural credit bank.

SCAER was already de facto abolished in the mid-eighties for lack of results. Only a few barns in some of the district capitals testify to the former existence of this state organisation. However the SOMIEX and OPAM, although they formally existed in the early nineties, during the research period, appeared to be non-existent in practice. The SOMIEX from its inauguration onwards could not respond to the people's demand for basic consumer goods such as sugar, salt and flour. In times of scarcity (times that for the SOMIEX returned with an increasing frequency) it was only in Bamako that the shelves were filled with the needed goods. In other shops the shelves were filled with -East German- razor blades and -Chinese- oil lamps. The SOMIEX -slowed down by red tape and mismanagement- simply could not compete with the private merchants who smuggled sugar, salt and flour into the country and sold these commodities (publicly). The liberalisation of the national commerce at the end of the eighties did not mean much more than a formal acknowledgement of what had existed already for over a decade in practice.

The persistent drought, used as an excuse by the Malian government for the malfunctioning of the OPAM in the seventies, may have affected the role of the OPAM. Furthermore after the drought period the impact of this organisation on the national commercialisation of agricultural products never reached an appreciable level. OPAM never succeeded in obtaining more than eight percent of the total commercialised millet production. Even in better years the organisation never succeeded in securing more than half of the total national market production, in spite of the fact that full control of the market production was its primary goal.

The already mentioned liberalisation in the case of the OPAM did not cause any substantial changes. Perhaps the biggest change was that the illegal sales of millet and maize (a general practice all over the country because of the low prices offered by the government) were legalised. OPAM was left with the following tasks: the regulation of the grain market through the sales and purchase of grains to stabilise the prices and the running of a cereal bank to hold buffer stocks. But the OPAM seemed unable to carry out the last task as it lacked the financial means to buy sufficient amounts of grain. Besides, international aid agencies have taken over most of the cereal bank projects.

The other group of economic services is formed by what are called the rural development services or in French *Opérations de Développement Rural* (ODRs). ODRs are regionally operating semi-governmental services that apart from their agricultural extension tasks have assumed more general socio-economic development activities.

Where areas under ODR jurisdiction is concerned, a clear spatial division can be observed between the Djenné District where four ODRs are operating and the rest of Central Mali where only one ODR is operational.[4]

The four ODRs - the *Opération Mils Mopti* (OMM), *Opération Riz Mopti* (ORM), *Opération Pêche Mopti* (OPM) and the *Opération de Développement de l'Elévage Mopti* - are active in Djenné because they consider the Mopti Region to which Djenné belongs as their working territory, hence the strict spatial division. The remainder comes under the responsibility of the CMDT, the *Compagnie Maliènne pour le Développement des Fibres Textiles*.

Table 4.2: Overview of main characteristics of the ODRs active in Central Mali

ODR	Main 'product'	Main sponsors	Territory
ODEM	cattle	World Bank	Djenné District
OMM	millet	USAID	Djenné District
OPM	fish	European Development Fund	Djenné District
ORM	rice	World Bank	Djenné District
		European Development Fund	
CMDT	cotton	World Bank	Koutiala District
		European Development Fund	San District
		CFDT	Tominian District

The ODRs although differing in product orientation and sponsoring (see also table 2) display many similarities in their geographical set-up, which approximates the administrative division to a large extent. Their territory is divided into *Secteurs de Développement Rural* (SDRs) which mostly coincide with districts and each SDR consists of *Zones d'Expansion Rurale* (ZERs), the counterpart of sub-districts. Every ZER is subdivided into clusters of five to ten villages (called *secteurs de base*)[5] whereby an extension officer (the *moniteur* or *encadreur*) is in direct contact with the rural population.

Finally three vocational training centres are found in the research area: the *Centre d'Animation Rurale*, the *Centre d'Orientation Pratique* and the *Centre Familial d'Animation Rurale*. These centres offer a similar package of mainly agricultural extension to a rather small group of young pilot farmers consisting of training in modern farming techniques and economic management of their farm enterprises. The impact of these centres in Central Mali however was nihil: in the entire duration of the project not a single pilot farmer was recorded in any of the surveys.

Social services
There are two major groups of social services: medical services and educational services.
In Mali, public health seems to be in a permanent critical state. The living conditions in the area, where malnutrition reaches critical levels in the weeks (and alas sometimes months) before harvest time, seem closely related to the low level of agricultural productivity. Malnutrition and poor sanitary circumstances result in many diseases which exact a heavy toll on the population, the infant mortality rates are estimated at 12 percent.

On both national and regional level, the task of the medical services, which except for some private pharmacies are organised by the central government in Bamako, is immense.

The geographical distribution of medical services closely follows the administrative structure. A strong concentration of specialised medical services is found in Bamako: there are five (out of fourteen) hospitals in the capital and more than half of all practitioners live and work there. The remainder of the larger hospitals are found in the regional capitals whereas smaller general hospitals (providing a restricted range of only polyclinic services to their patients) are available at district level. Also maternity clinics and pharmacies are available at district level. At sub-district level there are medical field posts where trained nurses (sometimes assisted by a maternity nurse) but most often only *bare foot doctors* provide medical services. Officially, polyclinic care is free of charge, although sometimes a small supplement (a few francs, a chicken) is required as payment for bandages, etceteras. Yearly vaccination campaigns are organised (mostly via schools) to prevent outbreaks of diphtheria, polio, tetanus, tuberculosis and measles. Both Catholic and Protestant missions are active in local health care. Near their missionary centres they finance and run polyclinics.

The distribution and sales of medicine is organised by the national pharmacy, the *Pharmacy Populaire*. This state organisation has pharmacies in Bamako, all regional and district capitals and in some of the larger sub-district capitals. In addition, one might find private pharmacies which are in fact "medicine shops" established by retired doctors and nurses to earn an additional income. The national pharmacy apparently is not able to meet the large demand for medicine. Ironically, the illegal sales of aspirins, malaria prophylactics and even antibiotics by private traders on weekly markets ease this demand.

The educational level of the Malian population is distinctly low. More than 90 percent of the active population has never been to school. Also in the near future this situation will probably not improve. The attendance rate (here defined as the percentage of pupils in the 7 - 14 years age group) is close to 15 percent in rural areas and approximately 30 percent in urban centres.

Formal education is based on the colonial French model. Primary education, compulsory but free of charge, is divided into two stages of respectively six (*premier cycle*) and three years (*second cycle*). At the secondary level, grammar schools (*lycées*) prepare the pupils for higher education as well as teacher training colleges and vocational schools (e.g. for agricultural and administrative training). Higher education is provided by colleges (called *Ecole Nationale*) for medicine, engineering and administrative (government) positions. To date, there are no universities. French is the medium of instruction although some schools experiment with local language.

Primary schools are widely spread throughout the country. In every administrative capital, sub-district capitals included, there are primary schools offering both *premier cycle* and *second cycle*. Also in a considerable number of non-administrative centres *premier cycle* schools are available, mostly in villages that once had a colonial administrative function. Secondary schools can be found in all regional capitals. Higher education appears to be a Bamako monopoly.

There are also private *premier cycle* and *second cycle* schools. These schools – mostly sponsored and managed by Catholic missions – offer an identical curriculum as their public counterparts and their teachers share a similar background in terms of training and qualifications. The difference between both types of schools is the school fee (10 to 20 US $

per semester) that has to be paid and the religious instruction. In addition to public and private schools is a third type of primary school, namely the *Ecole franco-arabe*, a public school where Islamic instruction is part of the curriculum.

A hierarchy of service centres

The strong connection between service supply and governmental administration is reflected in a hierarchy of service centres, which is virtually similar to the hierarchy of administrative centres. National head offices are located in the national capital of Bamako, and the further spatial dstribution of the services practically follows the administrative hierarchical lines; regional offices are found in regional capitals, district offices in district capitals and sub-district offices are located in the capitals of the *arrondissements*. The service centre hierarchy holds one more level at the base: formed by what we call "service villages". At this lower end villages can be found where, notwithstanding the absence of administrative services, a (albeit small) number of public services ranging from a primary school to a maternity clinic, a bare foot doctor or an agricultural extension service are available.

In the research area a three-tiered hierarchy can be recognised where four district capitals constitute the top layer, a score of sub-district capitals make up the middle layer and some fifty service villages form the base of the service hierarchy (see also table 4.3 and figure 4.3).

Table 4.3: A hierarchy of service centres in Central Mali

district capitals	
administrative services	district administrative services, police, court of justice, water and forestry (*Eaux et Forets*), post office
social services	secondary schools, (San, Koutiala), primary education, hospital, polyclinic(s), maternity clinic, pharmacy
economic services	district offices of ODRs, OPAM and SOMIEX.
sub-district capitals	
administrative services	sub-district administrative services, police post
social services	polyclinic, maternity clinic, primary schools
economic services	sub-district offices of ODRs
service villages	
social services	(occasionally) *premier cycle* primary school (occasionally) bare foot doctor, (occasionally) midwife
economic services	(occasionally) ODR extension officer

In order to obtain more insight into the geographical ordering of the service centres, all service centres were allotted a (theoretically defined) area of influence based on the centre's place in the above-defined hierarchy.

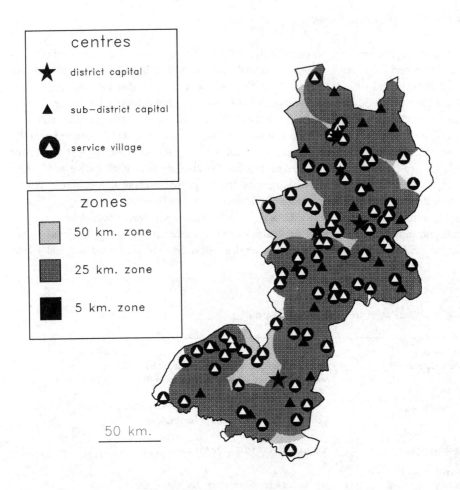

Figure 4.3: A spatial hierarchy of service centres in Central Mali

All district capitals were assigned an area of influence with a 50 km radius; 50 km is considered a good estimate of the maximum distance the population can cover (by public transport or bicycle)[6] in one day to reach e.g. the higher order services like the hospital. For sub-district centres an area of influence with a radius of 25 km is used. The rationale behind this distance is that the population may reach the (lower order) services located in this centre and return the same day.

From figure 4.3, showing both the hierarchy of centres and the outcome of the theoretical exercise, it is clear that the existing hierarchy at district and sub-district level offers ample possibilities to the population to visit the centres and use the services within the given distance limits.

Practically all parts of the research area seem "covered" and only minor parts on the periphery are outside the (theoretical) area of influence of the capitals.

At the lowest level of the service villages, a radius of five km seems appropriate because of the daily walk to and from a primary school. This distance also may be considered the maximum distance for pregnant women to travel in order to visit a maternity clinic. Looking at figure 4.3, however, even a 10 km radius does not provide sufficient coverage of Central Mali. A 15 km radius is required to obtain sufficient coverage and this radius may be considered too large for basic services such as primary schooling and basic health care.

All in all, however, it can be concluded that, considering the marginal socio-economic characteristics of this Sahelian country, in Central Mali the supply of services appears to be reasonably well developed and spatially evenly dispersed in the Pondo, the Transition zone as well as the Cotton zone. Up to this juncture, the supply side of services has been discussed. The demand side, of course, needs attention too.

The Use of Services

The use of services is measured in different ways. Counts of visitors were organised at a large number of service sites such as schools and medical services. Moreover, in a household survey a number of questions on the (household's) use of services were included. The surveys were conducted in the service centres Djenné, Sofara, Tominian, Koutiala and Zangasso and in a number of villages in their surroundings. Through this method information on service use could be analysed not only at the level of the research area but also at district and even sub-district level. This geographical differentiation made it possible to compare the functioning of services in the above-described agro-ecological zones as well. Here also, the services will be reviewed according to administrative, economic and social services.

The administrative services
The inventory of the use of administrative services already poses a problem. For in most cases these services are not really "used" by the population. The nature of administrative services is such that they "look after" the population and that direct contact between service and user only exists by way of exception rather than rule. The district *commandant* and his staff of officials, the police officers, the judge and his colleagues, all these civil servants work for the population without coming in direct contact with them. And if there are contacts, these come about through village leaders and village councils. Only for events such as wedding ceremonies, registration of birth, death etc. is a journey to administrative offices required.

Furthermore, there are situations where it is not the users who visit the service but rather the services which visit the users. The Water & Forestry officials visit the villages to check for forest fires (and to fine them heavily for offences, hence the unpopular image of this service).

Finally, the top-down character of the national administration makes it such that active participation in administrative and/or political matters remains rather restricted. Bamako sends down directives that hardly leave space for personal initiative. The administrators at district and sub-district level carry out these orders.

As to the "use" of political parties, the rural population does not "use" these services in any way. At district level, politics is purely an activity of a small group of well-to-do, urban-based, families. Furthermore, civil servants – mostly strangers to the villagers – "represent" the population in all kinds of democratic committees, such as the planning committee, the school boards, etc. During this research not one farmer was encountered in any of the existing administrative committees.

Active participation in co-operatives has proven to be restricted as well. There are only a few co-operatives which have more than 50 members (the cattle farmers' co-operatives in Djenné, San and Koutiala are the largest groups). As in the case of the political services, here also urban well-to-do families make up the lion's share of membership in the co-operatives. As a matter of fact, these co-operative structures do not really function for the envisaged target groups, rather they are used as tools by the richer (urban) families to profit from the financial support offered to the co-operatives (in Koutiala a civil servant co-operative has been established to profit from international food aid!).

In short, the co-operative movement is at the point of death, if any co-operative functions at all it is because of foreign financial support which helps to keep up appearances.

People make very little use of the postal services. This will certainly not come as a surprise, given that the population is relatively poor and hardly literate. Besides, the infrastructure the postal services offer is quite basic. Only in Koutiala and San are a limited number of houses connected to the telephone network whereas in Djenné only government buildings are in possession of a telephone line. Except for a hundred or so "ancien combattants" (war veterans, men who fought for the French in the First or Second World War) hardly anybody expects a postal cheque and the number of accounts at the savings bank facilities of the postal services tell their own story: in the centres of Sofara, Tominian, Djenné, and Koutiala respectively 30, 255, 155 and 489 bank accounts were noted. To make things even worse, more than 50% of the account holders had less than 4 US $ on their account whereas only 12% had more than 40 US $. Compared to the FCFA 45,000,000 (approx. US $ 180,000), the amount of wages paid by the CMDT to its employees' bank accounts every year, the total amount of savings at the Koutiala post office (US $ 32,000) seems of marginal importance.

As a general conclusion, it can be stated that people make marginal use of administrative services in Central Mali. On the one hand, this can be explained by the intrinsic nature of these services (in other words, there is no need to use these services). On the other hand, one might have expected a much more intensive use of some of the services we discussed here, such as the services provided by the co-operative movement.

The sectoral state organisations
The decline of the large state organisations has already been discussed. Therefore it is not surprising to find that, in the late eighties, the use of the services has been reduced to a bare minimum. The OPAM only plays its regulating role at the national level to control cereal prices. The SOMIEX shops are almost empty and often closed, its turnover per district shop (approx. FCFA 4,500,000 or US $ 18,000 per week) is comparable to those of a middle-sized private shop (larger private traders in Koutiala for instance reach turnovers of six to ten million FCFA.[7]

The SCAER has ceased to exist. The newly established national credit bank, the BNDA[8], did not have a branch in Tominian, Djenné and San at least until the nineties and the bank is present only in Koutiala-town and M'Pessoba (a sub-district capital near Koutiala). The most prominent reason hereof is the bank's intermediary function in the CMDT's farmers credit programme. It should be noted however that the people do not use the bank's services on their own accord. It is the CMDT which initiates the credit programme and invites farmers to participate.

In short, the sectoral state agencies have hardly anything to offer to the population and, although there is a need for services such as basic food products, basic consumer goods and credit facilities, the population does not need to use these service institutions to procure these services.

The rural development services
Although farmers or fishermen do sometimes visit the offices of the *Opérations de Développement Rural*, it is by no means a necessity. For the agents of the ODRs do visit their "clients" at home, in their villages. So, instead of counting "office visits", the intensity of contact between farmer (or fisherman) and extension agent needs to be examined. Most extension agents organise village meetings or channel their activities through village spokesmen, making the number of contacts a rather weak measure for "service use". In short, only indirect measures could be used to determine the intensity of service use.

First, a measure for the "density" of extension agents could be elaborated. This density measure is expressed in the number of agents per thousand households: the higher the "extension agent density" the more time agents may devote to each farmer and his family. The results are indeed surprising (table 4.4). The large differences between Koutiala district on the one hand and Tominian and Djenné on the other are clearly shown. In practice this means for example that in the districts of Tominian and Djenné one agent has to cover nine or ten villages whereas in Koutiala there is one agent for every six villages. In Tominian district, the recent changes in ODR organisation may have caused the low scores (the CMDT recently took over from the OACV and the ODR structures are still being reorganised), in Djenné this "excuse" does not apply. Four ODRs have been officially operating here for over ten years.

Also at sub-district level the same density measure could be determined. Generally speaking the same impressive differences exist. However within the Djenné District two sub-districts stand out: Konio and Mounia even have better figures than any of the sub-districts in Koutiala.

Table 4.4: The "extension agent density" per district and sub-district

	Extension agent density		Extension agent density
district		**district**	
Tominian	1.3	San	?
Djenné	1.5	Koutiala	2.3
sub-districts of Djenné		**sub-districts of Koutiala**	
Djenné	0.6	Koutiala	2.3
Konio	3.3	M'Pessoba	2.5
Taga	1.9	Molobala	2.2
Kouakourou	1.1	Kouniana	2.6
Mounia	3.1	Konseguela	2.2
Sofara	2.0	Zangasso	2.0

Apparently even within the territory of one ODR a differentiation in density is possible.

As a second measure, the acquaintance with an agent's name was tested. The outcome of this familiarity test shows a strong analogy with the agent density discussed above. In the sub-districts of Koutiala almost everybody (97 - 99 percent of all households) knew the agent by name whereas in Djenné and Tominian hardly more than 50 percent of the households could mention the agent's name. Most striking is the situation in the sub-district of Djenné where even less than 10 percent of the households could name an extension agent.

The diversity of rural activities in the sub-district of Djenné calls for some caution. It would not be fair to attach much importance to low "familiarity" rates of millet agents in a fishing village while an agent of the Opération Pêche cannot be expected to provide information on fisheries in a cattle raising community. Nevertheless, there are four ODRs operating in the district and one might expect at least one ODR agent to work in a village. In this respect, the considerably low score of 8 percent knowledge of the name of the fisheries agent in Sirimou, a true fishing village a few kilometres from Djenné-town, catches the eye.

As a third method used to get an insight into the use of ODR services, an attempt was made to determine where the rural population buys agricultural inputs. Again, a similar result is observed. In and around Koutiala and Zangasso 80-100 percent of the more than 1000 ploughs purchased after 1970 were directly bought from the CMDT. In the villages in the Djenné district, where ORM and OMM officially provide these tools, less than 10 percent of the ploughs were supplied by these two organisations and the majority of farmers bought their ploughs at the market. Also here, an inter-district differentiation can be observed: in the villages close to Djenné virtually no ploughs were purchased from the ODRs while around Sofara some 40 out of 400 ploughs bought since 1970 came from an ODR.

Despite the use of (inevitably) indirect methods, a number of very interesting conclusions regarding the use of ODRs can be drawn. To start with, the differentiation in use at the regional level is substantial. The only ODR in the Koutiala district, the CMDT, is intensively used by the inhabitants from all corners of the district. The use of ODRs in the Tominian district is restricted and reorganisation of the ODR activities are to be held responsible. The use of the four ODRs responsible for the extension services in the remainder of the research area, the Pondo, remains remarkably restricted. It should be added however that also within this area, at sub-district level, sizeable differences do occur and these differences seem not only to exist for geographically distinct areas, but also within one and the same sub-district where the use of one ODR may be much higher than the use of an other ODR operating in the same area. It goes without saying that these differences deserve our fullest attention in the final analysis.

Medical services

In a region where chronic diseases persistently threaten public health, one might expect every person to visit a polyclinic at least once per year. This would mean that the polyclinics in Sofara and Zangasso would have to cope with some 480 visitors every day. Such very high theoretical figures are not surprising in view of the meagre supply of basic medical services already discussed. In reality, nowhere in the region does the number of visitors even come close to these expected rates.[9] The rates of Tominian town are conspicious in the negative sense: its district clinic treats only nine persons per day.[10] Surprisingly, these low rates are found all over the region although missionary clinics seem to form the exception to the rule.

Table 4.5: A comparison of expected and actual numbers of patients seeking treatment at polyclinics in the research area

Service centre	Expected number of patients per week	Actual number of patients per week	Actual/expected x 100 %
district capitals			in percent
Koutiala	218	120	55
Djenné	339	130	38
Tominian	211	65	30
sub-district capitals			
Zangasso (K)	480	30	6
Sofara (D)	484	75	16
Mandiakui * (T)	156	750	480
Fangasso (T)	135	150	111
Timissa (T)	255	10	4
village centres			
Karangasso * (K)	218	210	96
Sanekui * (T)	156	250	160

K = Koutiala district D = Djenné district T = Tominian district * = missionary clinic

The number of patients seeking treatment at both the Catholic missionary clinics (Karangasso and Sanekui) and the Protestant missionary clinic (Mandiakui) easily surpass those of other clinics, including the district capital clinics. Maternity clinics are characterised by even lower rates. None of these clinics have more than ten clients per week. Pharmacies are well patronised. Every family buys medicine on a weekly basis. In fact, pharmacies are considered part of everyday life: in a bakery you buy bread, and if you have a headache you simply buy malaria pills or antibiotics (without a prescription). Control on the sales of medicine seems imposed only in the Karangasso pharmacy, a missionary service.

From the foregoing, general conclusions on the use of medical services cannot be drawn. Huge differences exist between the services and these differences have got nothing to do with geographical factors but seem fully attributable to the organisation providing the service.

Schools
All over the region, school attendance rates are very low. In our field surveys pupils of secondary schools were seldom encountered and even pupils of "second cycle" classes were scarce. To obtain an insight into the "use" of schools the potential school population (total number of children between 5 and 15 years) has been calculated, and based on school counts the actual school attendance could be determined. Both figures are related to the availability of classrooms per sub-district, the results are shown in figure 4.4.

It is clear that, if all children of school-going age indeed would attend classes, the school buildings would be overcrowded. And as figure 4.4 shows, especially in the spatially marginal sub-districts like Sy (on the west side of the Bani River) and Kassorola this crowding would be most felt whereas in the service centres constituting the top of the hierarchy the pressure on the schools would be less (indicating a relatively more developed school infrastructure). Nevertheless, more than 100 pupils would fill every classroom even with the most positively biased calculation.

In reality overcrowding of classes was not detected and the biggest pressure seems to be on the schools in the district centres whereas in the schools in the marginal centres classes are quite empty. The general impression discussed above is underlined by the data shown in Table 4.6. Most classes have a reasonable size and -again- only in the first classes of the primary schools in the higher-level service centres such as Koutiala and San is there a definite problem of overcrowding.

Finally, for a number of centres the school attendance ratio could be calculated at centre level. Except for Tominian (where of all children between 5 and 15 years, 67 percent actually attended school) all school centres scored rather low: between 23 and 38 percent. The meagre attendance ratio in all other villages (less than 10 percent!) reinforces the conclusion that the "use" of the schools indeed is rather restricted. The inadequate educational infrastructure apparently does not lead to frictions on the demand side.

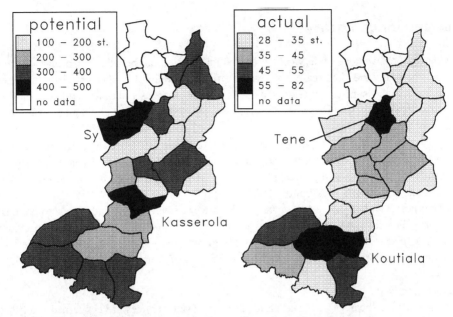

Figure 4.4: Potential and actual number of pupils per class

Table 4.6: Average number of pupils per class in primary schools in the research area.

	average per class+		average per class+
district centres			
premier cycle Djenné	43	*second cycle* Sofara (D)	33
premier cycle Koutiala	87-41	*second cycle* Zangasso (K)	12-25
premier cycle San	84-41	*second cycle* Mandiakui* (T)	29
premier cycle Tominian	43		
		service villages	
second cycle Djenné	39	*premier cycle* Senoussa (D)	50-25
second cycle Koutiala	54-51	*premier cycle* Kéké (D)	40
second cycle San	39	*premier cycle* Madiama (D)	52-18
second cycle Tominian	27	*premier cycle* Gomitogo (D)	80-16
		premier cycle Sadian (T)	30
sub-district centres		*premier cycle* Sanekui* (T)	33
premier cycle Sofara (D)	48	*premier cycle* Waratama* (T)	16
premier cycle Taga (D)	60-31	*premier cycle* Makoyna (T)	34
premier cycle Zangasso (K)	58-31	*premier cycle* Karangasso* (K)	52-19
premier cycle Mandiakui* (T)	50	*premier cycle* Sirakélé (K)	50-19

+ Where possible average figures for first and last class are given.
  * = Missionary school; D = Djenné District; T = Tominian District; K = Koutiala District.

An overview of service use

Drawing up the inventory of service use in Central Mali, it can be established that – despite the existence of a varied and reasonably well distributed supply of public services – the actual service use remains suprisingly restricted. Most administrative services are not used, nor are the sectoral state agencies. These latter services simply no longer have anything to offer. Also the use of social services can be considered restricted, especially when comparing the potential and the actual use at village level. Turning to the ODRs, a clear differentiation is observed: the use of CMDT services in the southern parts of the region can be considered remarkably high whereas the use of the four ODR´s operating in the northern parts sometimes hovers around rock bottom level.

Furthermore, within the northern part there are considerable local differences in ODR service use. For some of the public services the reasons for their restricted use are evident and already discussed above. For three groups of services –schools, medical services and agricultural extension services– a more detailed analysis of their (lack of) use is required.

Service use in detail

Where schools, medical services and agricultural extension services are concerned – all three groups are characterised by a hierarchically well-balanced distribution - one might have expected a much more intensive service use. At this point, a methodological sidestep into the foundations underlying the so-called Rural Centre Planning strategy seems appropriate. The core reasoning behind Rural Centre Planning (RCP) is the increase of the accessibility of services through the development of a hierarchically ordered network of service sites. Adopting Christaller´s logic on service location it is believed that the location of services within the reach of the rural population will indeed ensure the use of these services and, through this, the rural development process will profit (for a more detailed discussion, see i.a. van Teeffelen, 1992).

In the construction of RCP models, much attention is paid to the establishment of the radius of the service's area of influence. The problem of **service use** is as it were translated into a problem of overcoming **distance**, a reduction with, as will be argued further on, disquieting features.

A second assumption in RCP modelling is the (mostly implicit) notion that all inhabitants in the area of influence of a given service indeed will be able or willing to use the service. Although Christaller originally explicitly spelt out the implications of this assumption (and distortion of reality), modern development planners increasingly tend to forget this aspect and recent RCP models leave little space for socio-economic differentiation as a factor influencing service use.

A third assumption concerns the nature and the quality of a service. For the sake of convenience, it is assumed that services belonging to a certain type are of the same quality, thus, every dispensary is alike, all primary schools offer the same curriculum to their pupils and every extension officer offers the same product. Such generalisations obviously may collide with reality. Until this point, also in this study, these three assumptions were implicitly applied. Perhaps an explicit analysis on the relationship between service use on the one hand and distance, socio-economic differentiation and quality of services on the other offers a better insight into the processes underlying service use in Central Mali.

Distance and Service Use
For primary schools the relationship between distance and school attendance has been analysed and the results are shown in figure 4.5.
A very strong relationship appears to exist at (sub-) regional level and also at local level in larger district centres as well as in smaller –lower level- services centres. The results of the counts held at school level are fully consistent. The pupil population has a very high local component (80% and sometimes even more than 90%) whereas the number of pupils from surrounding villages remains low. At this point, it is useful to return to figure 4.3 where circles with a radius of 5-15 km have been drawn around service centres to provide an indication of the area served by school in the region. On this basis we now can conclude that even circles with a 5 km radius do not reflect reality in Central Mali. A radius of 3 km even 2 km would have been much closer to reality. A consequence hereof, of course, is that the total serviced area shrinks to a fraction of the estimated service area in Figure 4.3.

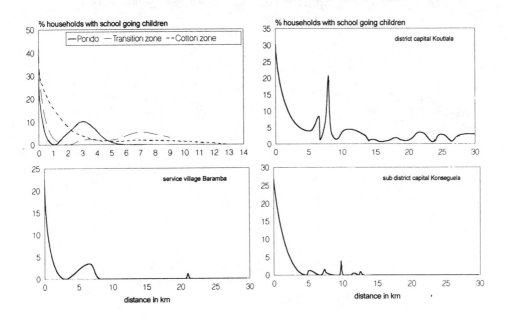

Figure 4.5: Distance and schooling related, results at regional level and some local examples

Patient counts at medical service sites show that here also a substantial part of the visitors is of local origin (50-75 %). In Koutiala town the percentage of local clients of the pharmacy even reaches 90%. However, the situation for missionary medical services in Karangasso, Sanekui and Mandiakui is clearly different. Here not only is the number of patients considerably higher (for instance 210 in the small village of Karangaso against 120 in the large centre of Koutiala town), but also the non-local patient component is relatively larger: in Karangasso over 75% of the patients are non-local.

Again, the hierarchical level of the service centres does not seem to play any role in the intensity of service use and the division of local and non-local visitors. Once again, the special feature of missionary services stands out: their clinics located in centres at the base of the service centre hierarchy attract more patients than medical services such as Koutiala and Djenné, the top centres in the region.

There is yet one other feature of this medical service use that surely requires comment. In figure 4.6 the relationship between intensity of medical service use and the distance factor is reflected in a curve. When several of these 'service use' relationships are compared in a graph it can be observed that all curves have a similar form but that there is a distinct differentiation in the degree of service use intensity per service.

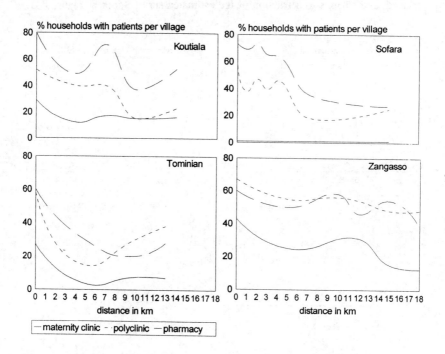

Figure 4.6: Use of medical services related to distance,
a comparison of four service centres

It is furthermore striking that this differentiation not only exists between different types of health care (maternity clinics having the lowest intensity) but also within one type of medical services, with maternity clinics showing the largest differentiation. The maternity clinic in Sofara is hardly visited whereas the Zangasso maternity clinic is frequented by a reasonable number of women and what is important the clinic also has a significant clientele living outside the service centre.

Furthermore, in higher order centres such as Koutiala and Tominian the 'service use' curves do not automatically represent a more intensive service use than those in lower order centres. On the contrary, the 'service use' curves for the polyclinic and maternity clinic in Zangasso indicate an even stronger impact on the hinterland than those of comparable services in the district capitals, and the Sofara pharmacy appears to have a much larger are of influence than the pharmacy in Tominian. This latter fact is all the more remarkable when related to the extremely low geographical influence of Sofara's maternity clinic. Obviously, an explicit distinction needs to be made between spatial use patterns of medical services, within the same type of service and also within the same service centre. Irrespective of how 'service use' curves differ in intensity, the form of the curves - the spatial trend in service use - seems to remain surprisingly identical. In other words, the curves seem to simply move up or down in the graph, thereby respectively enlarging or restricting the geographical extent of the service's hinterland.

Returning to figure 4.3, it can now be concluded that the circles drawn as hypothetical indications for the area of influence of medical services need not have the same radius. When it is clear that even medical services in one service centre can be characterised by differing areas of influence, a mix of smaller and larger circles should be drawn to indicate the area of influence of the respective medical services in the one service centre.

Finally, the manner by which agricultural extension services is provided is the opposite of the other services, that is, the farmers do not visit the extension officer but the officers visit the farmers in their village. The distance factor in this case is hardly comparable with the distance factor in the case of medical and educational services. Field research in Central Mali points out that although at regional and sub-regional level a distinct differentiation in service use exists, at the local level (within the area under the supervision of an officer) no differentiation at all could be observed. In terms of 'service use' curves this means that all curves should be drawn as horizontal lines at a certain intensity level. And - as in the case of medical services - the differentiation between services is reflected in a position higher or lower vis-à-vis the Y-axis.

Socio-economic differentiation and service use
Based on the household survey, insight into a number of specific socio-economic characteristics of the households in the research area could be obtained. In theory there are a large number of indicators in this field. In the practice of West Africa however, measuring socio-economic aspects can be a difficult and laborious undertaking. In this study three variables are taken into consideration: income, occupation and education.[11]

When comparing occupation with service use a rather vague result is obtained as shown in table 4.7. Perhaps farmers use services to a lesser extent than people belonging to other occupational clusters and perhaps government employees make more use of public services but a distinct differentiation does not exist and statistically no evidence for differentiation in service use could be found.

Table 4.7: Service use per occupation cluster (in % households per cluster)

Occupational cluster	Schools	Polyclinics	Maternity Clinics	Pharmacies
agriculture	65	39	24	52
government	85	66	36	85
trade	80	58	26	63
craft	53	45	39	70

For a number of medical services, this analysis could be repeated at service site level using service counts. But also at this very local level occupational differences in service use could not be detected. Also a possible relationship between income and service use has been tested. In this study income has been measured through a number of indicators including measures for durable consumer goods such as television sets, bicycles, motorcycles, and also ploughs, carts and oxen have been taken into consideration. However no proof whatsoever for any relationship between income and service use could be found. Perhaps the most intriguing outcome here is that also between the 'rich' Cotton Zone in the south and the distinctly poorer northern agro-ecological zones no differentiation in service use could be detected.

The educational level of the households has been measured in four ordinal classes. A "0-education" class comprising all heads of household without any formal education; a "1-6 years" class for heads of household with a maximum of 6 years (primary schooling) education; a "7-9 years" class comprising all heads of household with 7 to 9 years of education (9 years is equivalent to a complete secondary education); and finally a ">9 years" class for heads of households with post-secondary schooling.

In the case of medical services it turned out that the level of education does not seem to influence the intensity of service use very much. However, the same cannot be said of the results of this analysis for educational services! As Figure 4.7 also shows, the level of education of the head of the household directly influences the level of schooling of his children.

A closer examination of figure 4.7 even points out that the level of education hardly plays a role but that as soon as a head of household has any education his readiness to send his children to a school directly increases. The 'added value' of more education in this respect is very restricted: the distinction between non-educated fathers and educated fathers is important, not the level of education.

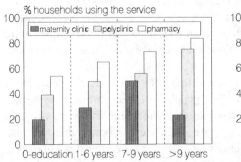 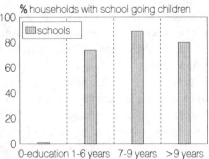

Figure 4.7: The use of medical services and schools related to the educational level of the head of households

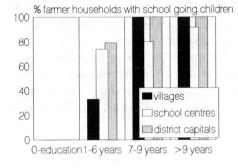 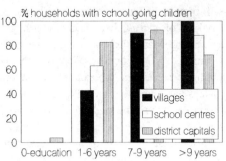

Figure 4.8: School use related to educational level of the head of household, according to occupational and residential aspects of the household

Figure 4.8 clearly shows that education regardless of the occupation of the father plays an important role: farmer households do not differ from other households in their behaviour, in other words when it comes to providing schooling for their children, educated farmers act like their counterparts in government service.

A second intriguing conclusion that can be drawn from the figure 4.8 is that the distance factor –amazingly– hardly seems to influence the schooling trends: the place of residence does not influence the father's decision to send his children to a particular school. We are faced with a contradiction. For on the one hand it has been concluded that distance does influence the use of schools, figure 4.5 seemed quite convincing. On the other hand figure 4.8 shows us the opposite.

The explanation for this problem is to be found in an obvious socio-economic aspect of Malian society: the geographical distribution of the population. It goes without saying that the lion's share of government employees reside where they work, e.g. in the service centres. Government employees, as a matter of course, are much more educated than for instance rural farmers. Therefore because of the larger concentration of the educated population in the service centres the tendency to send children to school will be much higher there as well, hence the high local peak shown by the school centres.[12]

Service quality and service use

In the small remote village of Karangasso both the medical clinic and the pharmacy succeed in attracting many customers not only from the centre itself but also from a large hinterland. The reason is to be found in the quality of the services offered: proper medicine is available, doors are never closed, the medical staff are capable and certainly do not expect small 'presents', etc. What a difference with many other medical centres where lethargy rules and the availability of medicine is more the exception than the rule.

It will remain difficult to operationalise the concept of "quality". Here quality is related to the certainty of getting help. Quality has got nothing to do with free treatment for in Karangasso and Mandiakui (both high quality missionary health centres) patients are charged for treatment and medicine and this does not prevent large groups of villagers from frequenting these centres. An example of the opposite end of the quality scale is easy to find: on Sundays large groups of villagers from faraway places visit Tominian because of the weekly market there. The medical centre is closed however because Sunday is a Christian day of rest.

The differentiation in use of agricultural extension services is also strongly related to this quality aspect. Where the ODRs succeed in adjusting their service supply to the needs of the farmers, a high utilisation occurs. In the villages in the Cotton Zone, without exception, farmers use the possibilities offered by the CMDT. Their service is widely appreciated and well known and CMDT meetings attract large crowds. The farmers in the Pondo clearly make much less use of the OMM, ORM, ODEM and OPM services.

It has already been discussed that the logistics and infrastructure of these northern ODRs are defective and surely these factors add to the malfunctioning of the services. This cannot explain however the distinct differences in service use intensity at the local level. Although the analysis of the geographical differentiation in service use is based on sometimes incomplete information, the outcome of the analysis has strong similarities with the analysis of the medical services. An ingenious and straightforward cause for these local differences in service use is the ineffectiveness of the extension officer himself.

To illustrate this we take the case of Sirimou, a village 11 km from Djenné. Only 7 percent of the heads of households can recall the extension officer's name. The friendly man can hardly be blamed for this. He does not speak the language of the Bozo population and out of frustration -because his ODR could not fulfil the villagers' demand for fertiliser and pesticides- he left the village and spent his days in Djenné playing soccer in the town's squad.

Small urban centres: at whose service?

Arriving at the close of the analysis of supply and demand of services in Central Mali, it will be clear that a simple yes or no cannot answer the above question. After all the first lesson to be drawn from the analysis is that different factors influence the functioning of the services under study and that -as a consequence- the nature of the service use can differ per service type and also per service site. Distance, in many development strategies -especially in Rural Centre Planning- considered as the dominant factor has proved to be only one out of several factors and certainly not the most dominant.

Socio-economic factors can have a strong influence on (the differentiation of) service use as has been shown for the educational services (cf. Ingram 1971; Moseley 1979). Even small differences in educational level lead to a large differentiation in the use of schools. Although it will not be easy to gain insight into these socio-economic differences, it proves to be indispensable to put a lot of effort into what Rushton called " the complexity of human behaviour" (1984, p.236). And besides, was it so surprising to find out that especially educated parents tend to send their children to school?

Also the quality factor deserves attention. It has been shown how quality differences in service supply lead to large differences in service use. For ODRs the quality factor should even be considered the dominant factor for service use. And also in the case of medical services the quality factor had a direct impact on the area of influence of the services offered. Not only quality, but also the nature of the services should be taken into consideration. In an area where the struggle for existence is a daily preoccupation and where good health is considered a luxury, post offices and in some cases maternity clinics do not enjoy the patronage of a large user group. The question which arises is whether it is worthwhile investing in these services while other basic services are still deficient. For what is the use of a maternity clinic like the one in Taga near Djenné where only chickens come to lay eggs?

However attractive it may be to 'mathematically' develop a Christaller-based location/allocation model for rural development, reality (or should I say the human factor) cannot be confined to circles. In Rushton´s words (1984 p. 236):

"...given the complexity of human behaviour it should come as no surprise that astute decision makers who contemplate the new location pattern of facilities recommended as the outcome of a location-allocation analysis, will often conclude that the given elements would not improve the system of services..."

Notes

1. I would like to thank Gina Rosario for her constructive assistance in improving my text, a smashing service!
2. The research project PROUST (Projet de Recherche de l'Utilisation des services et des Techniques) is the topic of my published Ph.D. thesis: *Dienstencentra en rurale ontwikkeling* (van Teeffelen, 1991).
3. Without entering into a protracted discussion, it can be stated that as products of the socialist plan economy these organizations are to be considered government services in many ways: they are managed by the government servants and manned by government staff and imposed on the population without any popular support.
4. Until 1981 the Tominian District was part of the territory of the 'Opération d'Arachide et des Cultures Vivrières' (OACV), an ODR focussing on the improvement of the groundnut production. The OACV did not function and its are of influence in Tominian was added to the CMDT territory in the mid-eighties.
5. The ORM, because of its focus on polders, uses a slightly different system in which clusters of polders are baptized 'casiers'.
6. In general, public transport by way of taxi *brousse* (old Peugeots with three rows of seats carrying a maximum of ten persons) or *baché* (pick-up trucks carrying approx. 25 passengers) is only used for inter-regional transport, sometimes *bachés* are used to travel to larger markets. In these cases 50 km daily return rides seem to be the maximum as well.
7. PROUST research published by Martin 1989, pp. 32-33.
8. BNDA: Banque Nationale de Développement Agricole.
9. Poor attendance at the official health facilities does not mean that the local population does not seek medical advice. In many cases, the traditional healers' advice is sought. In every village we find traditional healers and midwives, sometimes more than one. Some of them are true herb specialists, and administer effective treatment for a very small fee. Some are specialists in a few diseases, other are 'general practioners'. Of course we find plenty of witchcraft and suggestion, but it is a fact that the traditional healer has the confidence and the patronage of the villagers.
10. For the 150,000 inhabitants of the district there is only one practitioner who lives outside the district and visits the clinic irregularly.
11. The inventory of socio-economic characteristics of households was accompanied by various practical problems. These problems are related to West African culture. For instance 'occupation' is not given fact, most economically active men have more than one job, farmers are traders at the same time etc. Furthermore in the context of the research area income is in fact hardly measurable and can certainly not be expressed in terms of money let alone in e.g. monthly wages so that indirect indicators have to be used. For a detailed description of the operationalisation, see Van Teeffelen 1991, chapter 6.
12. Also in a statistical sense there is a strong relationship; from the LOGIT analysis it turned out that the educational level was responsible for 65 percent of the accounted variance, whereas the distance factor could explain less than 15 percent.

CHAPTER FIVE
The Absorption of Migrants and Their Living Conditions in Two Swaziland Towns

Cor van der Post

Rapid urbanization is characteristic of most developing countries during their post-colonial phase and the countries in Southern Africa are no exception. Thus, in Botswana, the urban population increased from 9.5 percent of the national population in 1971 to 17.7 percent in 1981 (CSO 1987). In Zambia, the increase was from 30.4 percent in 1969 to 40.4 percent in 1979 and in Swaziland from 14.2 percent in 1966 to 17.8 percent in 1986 (Swaziland Government 1966, Census 1986).

A disproportionate share of this urban growth however, has been the result of migration to the primate (capital) cities of these countries, e.g. Harare grew from 280,000 in 1960 to 654,000 in 1980 (O'Connor 1983) and Gaborone, in an even more extreme way from 3,812 in 1964 to 59,659 in 1981 i.e. from 0.74 to 6.34 percent of the national population (CSO 1987).

Smaller towns have, on the whole, exhibited a lower growth rate while stagnation is not uncommon. This raises two issues of concern: Firstly, the negative effects of over-urbanization (Hansen 1971) as evident from increasing unemployment, poor housing conditions and poverty in the large towns. Secondly, the concomitant lack of access to urban based services (Rondinelli 1985) for those living in rural areas far away from the larger cities, in which, together with the population, modern facilities have become concentrated. It also raises theoretical questions about the growth of and migration to small centres, as well as the practical issue of planning for such growth, which is considered desirable in many countries (see Mabogunje 1980; Rondinelli 1983).

This chapter uses data from Swaziland to examine the absorption of migrants in small towns in comparison with larger towns, with a view to highlighting similarities and differences, the understanding of which may contribute to more adequate planning for the growth of small centres. The absorption of migrants into town life is studied at two levels: their absorption into the urban labour force (i.e. economic absorption) and their absorption into the social life of the town (i.e. social absorption).

Labour-Migration and Rural-Urban Migration

Until fairly recently, labour-migration, especially to South Africa, was the major mode of mobility of the population of Swaziland and countries such as Lesotho, Botswana, Mozambique and Malawi. But in recent decades internal rural-urban migration to the larger towns has increasingly taken its place. This is, in Swaziland's case, evident from a stabilization of mine-labour recruitment for South-Africa and from the high growth rates of the towns in the country.

Existing models and theories of rural-urban migration (Lee 1966; Zelinsky 1971) as well as those of labour migration (see Amin 1974, Meilink 1978 and Forbes 1984, for a summary) do provide adequate descriptions of patterns (Amin 1974) and even of sequences of patterns (Zelinsky 1971), but do not offer a relevant framework for the analysis of changing patterns of migration in the less developed countries nor for differences in migration characteristics between large and small towns.

In order to provide such a framework, we may turn to a study of the interaction between the economic and the social dimensions of urban life and the different rates of change in each of them. Woods (1982) has remarked in connection with the explanation of mortality and fertility changes in modernizing societies, that such changes (as part and component of the modernization process) should be considered from two dimensions, i.e. the economic and the social.

With reference to migration it may be argued along these lines, that while the economic factor (i.c. capitalist penetration) engenders spatial concentration effects with resulting patterns of labour- migration to areas of capitalist accumulation and away from areas of disarticulation (Forbes 1984), feed back mechanisms exist in the social dimension.

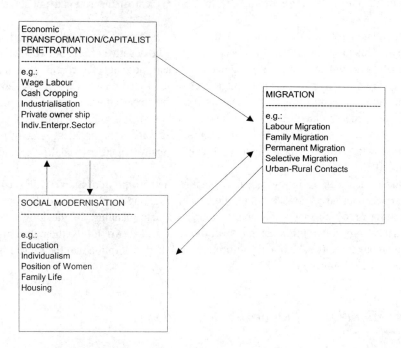

Figure 5.1: Economic and social dimensions of migration

The point is, that the social and individual behaviour of people is not fixed once and for all, but is subject to change as a result of what we may call social modernization. In this, education plays an important role, but also the gradual acceptance amongst the population of western culture (as expressed in dress, music, food habits, housing, marriage patterns) as well as western consumption patterns, which, although adapted to local traditions and conventions, result in new cultural and behavioral patterns. To become a permanent urbanite is, for many, part of this transformation process. The transition from temporary labour-migration to more permanent rural-urban migration may be explained by this dialectic between economic modernization (capitalist penetration) and the social dimension of modernization. The idea is summarized in figure 5.1. Temporary labour migration becomes socially less attractive as a result of changing patterns of behaviour (in themselves partly the result of economic modernization), in particular in relation to the changing role of women and changing marriage patterns (which in Swaziland implies a trend away from polygamy and extended families) and through rising levels of education. Differences in migration characteristics and in the absorption of migrants between large and small towns may also be studied in the context of this framework.

Social and Economic Absorption

In order to redirect urban growth to small towns, they need to become places in which migrants can be absorbed successfully. This requires both their economic absorption into the labour force (i.e. they need jobs and income) and their social absorption into town life. The aspects of social absorption examined in this paper include: the composition of migrant's households, their housing conditions and the extent to which they maintain urban-rural contacts after migration. The underlying assumptions are that increased levels of social absorption are associated with changes in the household composition of the migrants, (notably from the typical labour-migrants single person household to households built around families with children), with improvements in the housing conditions (Turner 1968) and with a reduced intensity of urban-rural contacts (Forbes 1984).

In order to assess the successfulness of the economic absorption of migrants, the absorption model of Titus (1985), based on Friedmann & Sullivan's (1974) model of the urban economy, was adapted to suit the conditions in Swaziland. Friedmann & Sullivan (1974) divide the urban economy into 'three major employment sectors distinguished according to the organizational form of activities within each sector' (p. 387). These are, arranged hierarchically in an ascending scale of labour productivity, economic power and social status: the individual enterprise sector of unemployed and self-employed workers that make up the street economy of a town; the family enterprise sector consisting of workers in small trade and service establishments with low capital to labour ratio; the corporate sector which includes workers in corporate enterprises as well as the government bureaucracy (for an alternative view with more emphasis on the capitalist and non-capitalist poles of the urban economy see McGee 1981). Each functions as a distinct subsystem of the urban economy and has specific absorption characteristics. In the corporate (or capitalist) sector absorption is mainly based on formal criteria, such as education, while family connections are more important in the family enterprise sector. The individual enterprise sector functions as a holding sector in which people wait for a chance to be absorbed in the other sectors. Absorption into the urban labour force leads to integration, but some migrants become more integrated than others. In Titus' model the degree of integration increases with the level of job security, the level of income and with commitment to the institutional order. In order to facilitate comparison, the model was adapted in minor ways. The corporate sector was split into a public and a private corporate sector to reflect the fact that, in Swaziland, government employees enjoy the highest level of job security. The individual enterprise sector was divided into the individual enterprise sector proper and the unemployed to bring out the low level of absorption of the unemployed more clearly. Within each sector a further subdivision was made on the basis of income, so that integration can be assessed along two scales: job security, as expressed in employment sector, and income. The third factor (institutional order) was not considered, as it could not be measured.

Income	Employment sector				
	Corporate Government	Corporate private	Family	Individual	Unemployed
High	6	5	4	3	-
Medium	5	4	3	2	-
Low	4	3	2	1	-
None	-	-	-	-	0

Source: Adapted from Titus 1985

Figure 5.2: Relative scores of integration levels by sector and income

A matrix was constructed with seven possible integration levels or scores, which partly overlap between the sectors, as illustrated in figure 5.2. The degree of integration is assumed to decrease from the corporate sector to the unemployed, on the basis of decreasing job security, and from high to low income. The model can thus be used to establish the success rate of migrant labour absorption, i.e. the degree to which migrants have been able to obtain access to jobs with good job security and income levels.

Thus, the highest integration level (6) is allocated to those with high incomes in the public corporate sector (government) because this sector combines a high degree of job security with a high income. The next level (5) is for those in government service with medium incomes as well as those employed by the private corporate sector with high incomes, while level 4 is for persons in government service with low incomes, private sector workers with medium incomes and family enterprise workers with high incomes (see figure 5.2), etc. Finally, the unemployed constitute the last category (0), as they are not (yet) absorbed into the urban labour force.

This model will be used to assess differences between economic absorption of migrants in large and small towns in Swaziland. It is assumed that absorption will be more difficult in the categories with high levels of integration due to more strict criteria for employment (e.g. education), and will be more easy in the lower order categories.

Background to the Study

The kingdom of Swaziland is the smallest of the Southern African countries in area (17,000 square km) as well as in population (676,000 in 1986). Urbanization is a recent phenomenon (even in 1966 there was not one town in the country with more than 20,000 inhabitants) and was closely associated with the colonization of the country.

When Swaziland became a British Protectorate in 1902, the effects of the penetration of capitalist modes of production and consumption on her (traditional) spatial structure were already becoming obvious. Amongst these was the alienation of almost 50 percent of the total land resources of the people by means of concessions, granted under dubious circumstances by the Swazi kings and confirmed as legal by the British authorities, to enterprising Europeans interested in mining and farming (Crush 1982). Secondly, the institution of labour-migration had long lasting social and spatial effects. Migrant labour was recruited from 1885 onwards, initially for the mines in Kimberley (Booth 1982) and later for those in the Witwatersrand as well as for European owned enterprises in Swaziland, giving rise to a flow of labour-migrants, which still continues today (in 1985 over 16,000 Swazi's migrated on this basis to South Africa).

By 1966 the country's spatial structure could be described in terms of economic cores with concentrated patterns of modern development lying as islands in a sea of underdeveloped periphery (Fair 1969) (see figure 5.3). Concomitantly urban development, being part of this development process, also occurred. Company towns were established for the workers of forestry plantations, mines and sugar and citrus estates. These towns are still legally not considered as towns, but are governed by large companies involved in the exploitation of the country's natural resources. Their role as regional centres therefore, remains very limited in most cases.

Administrative centres were established by the colonial authorities in each district, while Mbabane became the national capital. Some of these centres developed regional service functions and in some cases also industrial activities (Whittington 1971) which are mainly catering for the development core areas. In the recent past centres like Manzini and Mbabane have experienced an accumulation of functions and a concentration of population and consequently grew out to the largest towns, comprising 67 percent of the total urban population in 1986. Considerable growth also took place in a selected number of small centres, notably in Nhlangano, Piggs Peak and Simunye, but stagnation occurred in most company towns and other small centres. Consequently, an unbalanced urban system developed as is typical for most developing countries.

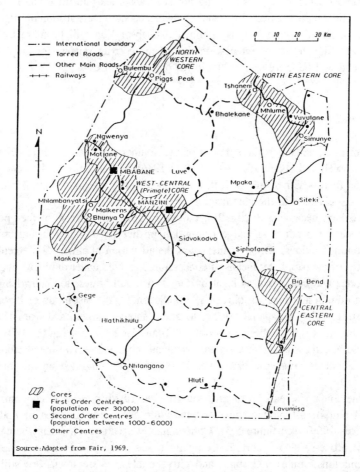

Figure 5.3: Swaziland's major economic cores and urban settlements in 1986

Presently, the urban settlements may be divided into three categories (see figure 5.3). First order centres include Manzini and Mbabane, both with over 30,000 inhabitants in 1986 and with a concentration of national and regional functions. Second order towns such as Nhlangano, Piggs Peak, Siteki, Tshaneni and Malkerns have populations between 1000 and 6000 and function as regional centres. There are administrative or company towns with regional functions. Third order centres are rather small and usually stagnating company towns or villages.

In spite of commendable efforts towards the development of settlement planning (with an emphasis on decentralisation) under the second and third National Development Plan and more recently the National Physical Development Plan (see Swaziland Government, no date (a), and 1986) very little has been achieved so far due to the poor implementation of plans and the absence of a legal framework for such planning.

Early migration to the towns was to a large extent of the labour migration type. In the case of company towns, this is still true and evident from their extremely unbalanced age and sex ratio's. But otherwise, internal migration has become increasingly permanent and more significant than international labour-migration.

Since the Second World War, the major rural-urban migrant flow has been directed towards the two large towns, Mbabane and Manzini. But there is also a flow to some of the smaller towns in the country. Of these, recently, only the larger administrative centres, such as Nhlangano and Piggs Peak, have experienced substantial growth. Most of the other, smaller ones, have stagnated and have not attracted any migrants.

The research for this paper was carried out between 1982 and 1984 and consisted of two independent urban surveys, one in Manzini, as an example of a large town, and one in Nhlangano, as an example of the few small towns (apart from newly established company towns) which had attracted migrants in the recent past.

In Manzini, 602 households (12.4 percent of the total) were selected in a random sample, stratified by residential area, and details were recorded of 1693 individuals over 14 years old (of whom 83 percent were migrants, i.e. persons not born in the town), concerning migration history, employment, income, education and housing. For most variables a confidence limit of five percent at the 95 percent confidence level is applicable. A similar survey was carried out in Nhlangano through a systematic sample of 131 households (12.1 percent of the total) with 283 resident individuals of 14 years and older, of whom 84 percent were migrants). The confidence limit for the 95 percent confidence level is in this case about ten percent for most variables.

Economic Structure and Employment in Manzini and Nhlangano

The town of Manzini, situated in the central part of the country, had a population of 43,000 in 1986, making it the most populous town in the country and ahead of Mbabane, the capital. Rapid growth characterized Manzini's recent history: its population was only 146 in 1911 and 16,106 in 1966. This growth was associated with its economic development and the town presently is the most important manufacturing centre in the country, in addition to being the administrative capital of the Manzini district and a major national and regional service and transportation centre. The employment opportunities thus are varied (see table 5.1) and many migrants are attracted to the town from all over the country (see figure 5.4).

Nhlangano, in the south of the country, is much smaller. In 1986 the population was 3,689, making it the sixth largest town in the country. Its population growth is fairly recent and as late as 1958 it still had less than a 1000 inhabitants. The economy is dominated by government activities in which 40 percent of the labour force finds employment, as it is the capital of the

Shiselweni district. Other important activities include manufacturing/crafts, retailing and other regional services (see table 5.1). This town attracts its migrants mainly from the Shiselweni district (see figure 5.5).

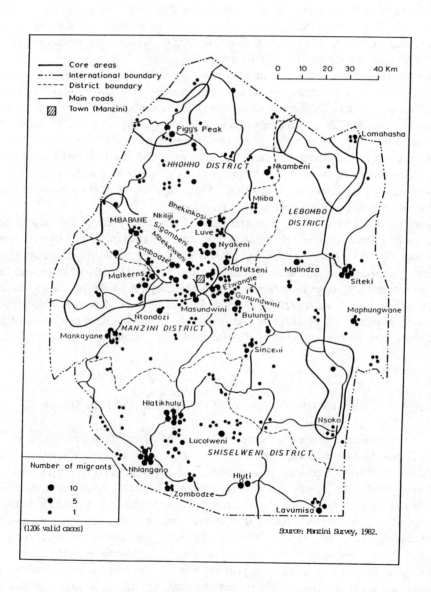

Figure 5.4: Place of birth of Swaziland born migrants living in Manzini, 1982

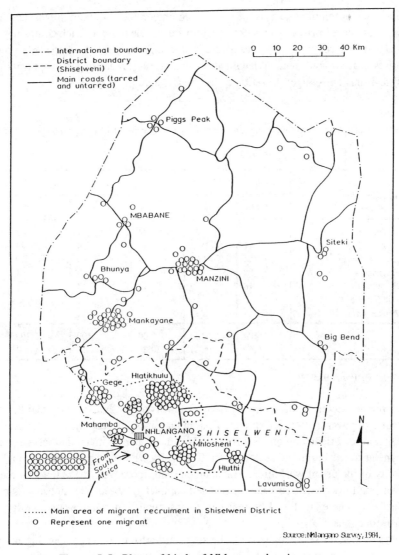

Figure 5.5: Place of birth of Nhlangano's migrants

The general employment rate is high in both towns, and the rate of unemployment thus relatively low. In Manzini 76 percent of the males over 14 years old are employed and 48 percent of the women; in Nhlangano a slightly lower 73 percent of the men and a slightly higher 52 percent of the women, resulting in a participation ratio of 0.34 in Manzini and 0.37 in Nhlangano. In both cases these figures are much more favourable than those of the surrounding districts (0.26 for the Manzini district and 0.14 for Shiselweni), emphasizing that towns are places where jobs can be found.

Corporate sector employment is of overriding importance in both towns and extends to relatively small enterprises. The background for this is that many of such companies are owned and operated by foreign nationals, or naturalized foreigners who are compelled to operate their enterprises in accordance with all laws and regulations (e.g. regarding the payment of minimum wages) and thus are obliged to set up rather formal enterprise types.

Table 5.1: Employment by type of economic activity and by Sex, Manzini and Nhlangano (in percent of the employed of 15 years and older)

Type of economic activity/industry	MANZINI			NHLANGANO		
	Males	Females	Total	Males	Females	Total
Government Service	25	23	24	49	29	40
Retail/Wholesale	12	17	14	14	23	18
Handicraft + Repair	3	14	7	3	9	6
Manufacturing	20	5	14	4	1	3
Services	18	9	15	14	11	12
Domestic Service	3	19	9	2	13	7
Other/Unknown	18	13	17	14	14	14
Total	100	100	100	100	100	100
N =	(609)	(430)	(1039)	(129)	(154)	(283)

Source: Van der Post, 1988

The corporate sector is of relatively larger dimension in Nhlangano, mainly due to the importance of government employment. This, in fact, is not in accordance with ideas prevailing in the literature. Friedmann and Sullivan (1974), for example, maintain that the proportion of employment in the corporate sector in Third World towns is relatively small, including only a minority of the urban workforce. Mabogunje (1980) moreover, maintains that the corporate sector in small towns will be very small if not completely absent. This clearly is not the case in Swaziland, where, it should be realized, towns small in population may be large in terms of urban functions.

The family enterprise sector tends to be small in both towns, encompassing not more than eleven and sixteen percent of the labour force, mostly domestic servants. The individual enterprise sector is also of relatively minor proportions, twelve percent of the labourforce in Nhlangano and ten percent in Manzini, while the proportions of the unemployed are 23 in Manzini and eleven percent in Nhlangano, respectively. The unexpected small size of the individual enterprise sectors may be partly explained by the high level of competition from formal sector shops and the restrictive licensing regulations for street-selling (ILO 1977).

Thus Nhlangano has a larger corporate sector, a smaller family enterprise sector and fewer unemployed than Manzini. The differences between the two towns however, are complicated by differences between the sex-structures of the labour force within each town. In a sense we are dealing therefore, with four rather than two employment structures (see figure 5.6).

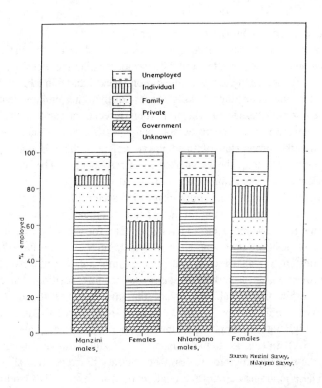

Figure 5.6: Employment structures of Manzini and Nglangano

These are a reflection of the different mix of production and servicing activities and the existing sex-selectivity of employment in each town. Certain jobs, for example manual jobs in construction and manufacturing, are accessible only to men, while other jobs, such as domestic services, are considered to be women's jobs. Other jobs demand high levels of educational achievement and are not discriminative according to gender at all.

Economic Absorption in Manzini and Nhlangano

The model of economic absorption was applied to the sample of migrants in the two towns in order to establish the degree and the level of absorption of migrants into the urban labour force. This refers to the questions whether migrants have been able to secure jobs for themselves and in which sections of the urban economy. Table 5.2 provides data on the level of overall employment and table 5.3 on the sectors of employment. Migrants are differentiated into those who migrated during the four years before the survey (recent migrants) and those who migrated earlier.

In Manzini, both male and female migrants enjoy higher employment ratio's than non-migrants and do not appear to experience any disadvantages in access to job possibilities (see table 5.2). Their favourable absorption record into the labour force is confirmed by their higher than average incomes compared to non-migrants.

The great majority of the migrants in Manzini have acquired access to relatively secure and well-paid employment in the private corporate sector, while the public corporate sector apparently is also easily accessible to them (integration levels 5 and 6 in table 5.3). This applies at least to men with some education or skills. Access to the other employment sectors either seems less favoured or more restricted. The proportion of recent migrants in the lowest position of the individual enterprise sector is considerably below that for more established migrants and non-migrants, while the proportion of unemployed amongst them is consequently higher, although not higher than among the established male respondents. The main reason for this being that most recent migrants simply cannot afford to stay idle.

For recently migrated women, the family-enterprise sector (domestic service) is of great significance and for many rural women the job of 'maid' is their first one in the town. Here unemployment rates for newcomers are much higher than for the longer established women, due to the dependent position of the most recent female migrants who have just joined their husbands or families. It was also found, that for most migrants the move to town implies a considerable improvement in employment status and income position. Once in the town, however, the economic and social mobility of the migrants was found to be extremely restricted, as has been confirmed by similar findings elsewhere (see Goldscheider 1983).

In Nhlangano the high rate of employment participation among migrants is comparable to Manzini (see table 5.2). In contrast to the situation in Manzini, however, the participation rate is highest among the recently migrated males, who contrary to expectation show low levels of unemployment. These migrants therefore, seem to experience especially favourable conditions for absorption into the urban economy.

Table 5.2: Proportion of the economically active population employed amongst various categories in Manzini and Nhlangano

Population Group	% Employed Manzini		% Employed Nhlangano	
	Males	Females	Males	Females
non-migrants	81	49	--	--
migrants	91	67	90	89
recent migrants	88	61	92	86
earlier migrants	93	73	88	91
total (incl. non-migrants)	89	64	87	90

Source: Van der Post, 1988

This is further confirmed by their higher than average income, not only at the personal level, but also at the household level. This pattern therefore, is rather a-typical. A large group of male migrants apparently enter the urban scene with one giant leap and establish themselves immediately at a high position in the urban employment hierarchy. The reason behind this is that quite a number of such migrants actually are experienced and well-educated government officials who have been transferred by their employer.

For women, the situation is quite different and resembles more the situation in Manzini; recent female migrants experience higher unemployment rates and lower incomes as compared with the more established migrants. While in Manzini the majority of the migrants find jobs in the corporate sector; in Nhlangano these are mainly government jobs (see table 5.3). This applies in particular to males, whereas for women the domestic services again play a key role in the absorption of recent migrants. And, again as in Manzini, there was no significant upward economic or social mobility after settlement in Nhlangano.

By adding the proportions in each integration category, a simple cumulative index can be calculated, with a highest possible value of 700 (i.e. 100% employed in category 6) and a minimum value of 100 (i.e. 100% employed in category 0). The overall level of integration of the various categories of the employed migrants in both towns can now be compared more directly by examining the relative scores of the cumulative integration index (see table 5.3). The implicit expectations are that the integration scores would be higher for more established migrants and lower for the more recent ones; that they would be higher in Manzini with its greater diversity of employment opportunities than in Nhlangano, and finally that males would score higher than females.

Table 5.3: Integration level of migrants in Manzini and Nhlangano (in % of economically active persons over 14 years) (*)

Integration level (+)	Male migrants				Female migrants			
	Manzini		Nhlangano		Manzini		Nhlangano	
	Pre-1978	1978-82	Pre-1980	1980-84	Pre-1978	1978-82	Pre-1980	1980-84
6 (GCh)	6	5	0	4	0	1	3	3
5 (GCm, PCh)	23	28	14	44	16	17	13	15
4 (GCl, PCm, FEh)	40	35	37	23	15	7	28	15
3 (PCl, FEm, IEh)	6	10	23	15	5	3	16	20
2 (FEl, IEm)	11	13	6	6	19	21	19	23
1 (IEl)	2	1	6	0	14	8	12	6
0 (Unemployed)	12	8	14	8	31	43	9	18
Total	100	100	100	100	100	100	100	100
Cumulative Integration Index	453	467	405	493	307	278	393	365
N =	282	222	35	48	280	250	32	34

(*) Category Unknown excluded from calculation
(+) Explanation of codes: G = government; P = private; C = corporate sector; F = family; I = individual; E = enterprise sector; h = high income; m = medium income; l = low income
Source: Van der Post, 1988

In all categories women score lower than men, because many of the better jobs are virtually reserved for men and thus not easily accessible to them. The exception is found in the jobs which require tertiary education; for these, women can compete successfully with men.

Amongst women, recent migrants score lower on the integration scale compared to more established ones, thus in accordance with the expected trend. But men do not behave according to expectation. While in Manzini earlier and more recent male migrants score almost equally, in Nhlangano the recent male migrants score much more favorably than the others, thus upsetting the pattern. The single most important reason lies in the recent transfer to that town of a relatively large number of highly educated and experienced government officials.

Contrary to expectation all categories in Nhlangano, except for the pre 1980 male migrants, score higher than their counterparts in Manzini. The difference for the women is largely explained by their lower rate of unemployment (partly caused by the fact that fewer women are looking for jobs outside the home), while for men the explanation is once more the government employment factor.

Thus, in both towns migrants have been able to secure jobs for themselves and have, in general, not experienced any disadvantages in access to jobs. This applies not only to well established migrants who became urbanites many years ago, but also to those who have only recently migrated to the towns. But, generally speaking, the job situation is far less favourable for the women in both towns and access to higher status positions, including those in the private and in the public corporate sector is much more limited for women than for men. However, this applies to migrants as well as to non-migrants. In fact, integration differences are more pronounced between the sexes than between different categories of migrants of the same sex. The evidence suggests that the economic absorption of migrants in small towns is not inferior to that in large towns and for certain groups may even be superior.

Social Absorption in Manzini and Nhlangano

The process of transformation from a rural inhabitant to an urbanite is a complex one and absorption into the urban labour force is only one aspect of the integration of migrants. Of further relevance are aspects of life in the social and psychological realms. In this paper we briefly review household composition, housing conditions and the connections of migrants with their rural origins as indicators of their social integration into town life.

Household Composition
The largest group (44.9 %) of households headed by migrants in Manzini, consist of nuclear families, i.e. husband and wife with or without children. In fact the majority of all migrant households are built around a nuclear family, with sometimes additional relatives, usually a brother or a sister, or occasionally non-relatives. This includes category 3 and most of category 4 in table 5.4, together comprising 60.6 percent of all migrant households in Manzini, while another proportion consists of incomplete nuclear families, i.e. usually a woman with children (type 2 in table 5.4).

Another important type of household is the single person household (10.3 percent of the total), in most cases a man living alone in a room, i.e. the typical labour migrant type of household. It is significant that this category is quite small in Manzini (in spite of a degree of underestimation due to the inclusion of some who have joined other households and here are classified in categories 5 or 4). The migrant household composition with its large share of complete households, thus is an indication that rural-urban migration today has reached a relative mature and permanent stage with the apparent intention of these households to establish themselves firmly in town.

Table 5.4: Composition of Migrant Households in Manzini and Nhlangano

Type of household	MANZINI (%)	NHLANGANO (%)
1. One person, Male or Female	10.3	29.5
2. One person with children	13.3	12.2
3. Married couple (with or without children)	44.9	32.2
4. As 2 or 3 above with other relatives or non-relatives	24.1	9.6
5. Other composition	6.8	15.7
6. Unknown	0.6	0.8
Total	100.0 (N = 534)	100.0 (N = 115)

Source: Van der Post, 1988

The composition of the migrant-headed households in Nhlangano is also shown in table 5.4. Remarkable is the much higher percentage of households consisting of one member only (29.5 percent). A correspondingly lower percentage consists of households with married couples and children and/or relatives (a total of 41.8 percent).

Of the single person households, the majority (21.7 percent of all households) has a family 'at home', in the rural areas. These are typical migrant-laborer households. It confirms that migration to Nhlangano has a smaller permanent component compared to Manzini. The above is supported by evidence concerning the changing population structures of the towns. Nhlangano's population structure - with 53.1 percent of its population of working age at the time of the survey, as against 49 percent in Manzini - deviated more from the national population structure than Manzini's. This is because in Nhlangano fewer workers have their children with them in town.

In addition, the population structure underwent more limited change in Nhlangano compared to Manzini. The change in the proportion working-age persons form 54.1 percent in 1976 to 53,1 percent in 1984 in Nhlangano, while more substantial change occurred in Manzini: from 55.5 percent in 1976 to 49.0 percent in 1982.

It can be added, that this pattern can be recognized in the motivations of the migrants: in Manzini 40 percent of the migrants were motivated directly by work, while 40 percent came as dependents. In Nhlangano only 25 percent came as dependents and 60 percent for work.

The Housing of Migrants
Theories about this aspect of the integration process of migrants suggest that the typical rural migrant arrives as a poor, unemployed person, who will first settle in a part of the town close to employment opportunities and cheap housing. With time, the migrant will generally improve his position by obtaining a better paid job and thus will experience a general process of upward social mobility which will also be reflected in his housing conditions (see Turner 1968).

Do migrants really improve their housing conditions with increased length of residence in town? There is in fact evidence of some form of mobility in this respect in Manzini (see table 5.5). House ownership correlates positively with length of residence. Among migrant heads of household who arrived since 1975, only 20 percent are house owners, while for migrants who arrived earlier this is 56 percent. This applies in particular to low income groups, while high income groups often remain tenants, as they are usually provided with subsidized good quality housing by their employers, (i.e. in many cases the government). There is, however, no difference in housing quality between the different categories of migrants, while it is clear from additional survey questions that few migrants make substantial improvements to their first house in the town.

In the case of Nhlangano, recently migrated households tend to occupy, on average, better houses than more established migrants. This is reflected both in the general quality of the house and in the facilities available, such as water (see table 5.5). The explanation for this apparent anomaly lies in the fact that a very large proportion of the recent migrants are government employees who upon arrival in the town are provided with accommodation by their employer. Such houses are provided at subsidized rents and generally are good quality.

Table 5.5: Housing situation of migrants in Nhlangano and Manzini, by period of migration (in % of heads of households)

	MANZINI		NHLANGANO	
	Migrants pre 1975	Migrants post 1975	Migrants pre 1980	Migrants post 1980
High quality house	40	40	51	72
Low quality house	60	60	49	28
House owned	56	20	34	4
House rented (or other)	44	80	66	96

Source: Van der Post, 1988

House ownership exhibits a more familiar pattern. The proportion of households owning their house increases with length of urban residence. Among recent migrants the proportion of house owners is very low (4 percent) and this again is a reflection of the fact that many are accommodated in employer (i.e government) provided housing.

Even amongst the more established migrant households, the proportion of house owners is relatively small (34 percent), as a result of the same fact (see table 5.5).

Thus, new migrants are better housed, while more established migrants are more often house owners. Government employment is such an important component of the total that it heavily influences the overall housing picture and perhaps neutralises trends which would otherwise be present. This becomes clear when it is realized that 49 percent of all the houses in the town are owned by the government, which is 63 percent of all rented houses.

Urban-Rural Contacts

Of special economic and social relevance are the connections which migrants continue to maintain with their home area, once they have become town dwellers. To what extent do they maintain access to their land, do they continue to keep a house in the rural home area, do they remit money home, or do they even keep their wife and children at home while working in the town? It may be argued that the degree to which migrants maintain such connections is illustrative of their degree of integration into the urban way of life. However, a complicating factor is that such contacts may also function as a form of investment for urbanites with some spare money (Bell 1986). Bearing that in mind, some differences between the two towns nevertheless are remarkable.

Table 5.6: Migrants connections with the home area, Manzini (A) and Nhlangano (B), by period of arrival (*)

A. MANZINI			
Period of Arrival	% Without Connections	% With house land or cattle	% Visiting Only
1920-1929	100	--	--
1930-1939	50	--	50
1940-1949	29	10	61
1950-1959	22	30	48
1960-1969	21	28	51
1970-1979	14	38	48
1975-1979	6	40	54
1980-1982	5	40	55
All Migrants	11	35	54

B. NHLANGANO		
	Pre 1980 migrants	Post 1980 migrants
Owns house at home	58	57
Owns cattle at home	32	24
Has land at home	42	27
None of the above	19	23
Proportion regarding Nhlangano as 'Home'	43	14

(*) Due to differences in the survey questions data for the two towns are not entirely comparable.
Source: Van der Post, 1988

In Manzini an inverse relationship could be established between the time migrants have spent in town and the intensity of their connections with the home area, in the form of ownership of a house, of cattle or access to land. About one third of all migrants maintained such connections, but among recent migrants the percentage was slightly higher, about 40 percent (see table 5.6A). Apparently, migrants in Manzini are keen to maintain connections with their home area in the early stages after migration, but with increasing urban residence tend to give up their house, land and cattle in the rural area.

In Nhlangano, such a trend could not be established. Although an increasing proportion of the migrants considered the town as 'home' with increasing length of residence, migrants in Nhlangano, in general, tended to maintain stronger connections with their 'home area' than those in Manzini (see table 5.6B). While in the latter towns, even among the most recent migrants only about 40 percent maintains a house (with or without land or cattle) 'at home', the proportion was 57 percent for all migrants in Nhlangano and almost equally high amongst the recent and more established migrants. Many have cattle and land in addition, but recent migrants less frequently than pre 1980 migrants (see table 5.6B).

In Nhlangano the majority of the migrants thus maintains a substantial foothold in the home area, which can be taken as a sign for their limited degree of integration into town life. Data on the remittances of the migrants in this town confirm this: among recent migrants, the percentage remitting as well as the amounts remitted tend to be higher than for earlier migrants, as could be expected. But the differences between the two groups were very small: non-remitters were 31% among the recent migrants and 39% amongst the pre 1980 migrants. The average amount remitted also differed little. Pre-1980 migrants remitted on average 36 E(malangeni), against 38 E per month for the most recent migrants. Such differences present only very weak evidence for the increased integration of the migrants with increasing length of urban residence.

Conclusion

Our evidence suggests that the economic absorption of migrants tends to proceed equally well (or better) in small towns as in larger ones. It may be argued, of course, that the large component of government employees in our case study of Nhlangano is a special factor. But it also is a general factor, as many of the small towns with growth potential apparently are government centres. This suggests that the government may play a crucial role in stimulating their development.

A substantial difference of course exists in the scale of migrant absorption. Manzini, as the largest town in Swaziland, annually absorbs more migrants than all small towns in the country combined; its annual absorption capacity is eight to ten times larger than that of Nhlangano. There is also a considerable contrast in social absorption between the two categories of towns, the evidence suggesting a lower degree of social absorption of migrants in small centres.

For a possible explanation of the differences between our two towns, we may return to figure 1, which considered the interaction between the economic and the social dimensions of urban life and the different rate of change in these. The rate of change is not only different between the dimensions, but also between spatial levels.

Certainly there are economic differences between the towns. Large towns offer a wide range of employment opportunities at different levels of skill and distributed over all employment sectors from the corporate to the individual enterprise sector. Notably, there are many jobs for people with little education, but also for those who are relatively highly educated. Comparatively, there is a large number of jobs in the family enterprise sector, which in

Swaziland must be regarded as a form of economic modernization because it involves jobs which did not exist in the traditional economy but emerged with the development of capitalism in the urban economy. Corresponding with this type of development there are also relatively many unemployed in the larger towns.

In small towns, like Nhlangano, the range of job opportunities is more narrow. There are, in particular, fewer jobs for the unskilled and for the highly skilled. Especially the opportunities in the family enterprise sector are more limited. That type of modern development obviously has not yet taken root in the small town economy. Moreover, employment rates among women are much higher, probably reflecting the persistence of the traditional African pattern rather than social and economic processes of modernization.

Differences in social modernization were also identified on the basis of evidence concerning the household composition, housing conditions and urban-rural contacts. What is significant in the larger town is that the data convey the impression that most migrants do not regard the town as just a place to work, but also as a place to live, to marry and to bring up and educate children. This is evident from the high proportion of workers who bring their families to town, the limited degree to which contacts with the rural areas are maintained after a few years of residence in town, and from the increased house ownership with length of urban residence.

In many small towns, a large proportion of the employed consists of relatively well-paid and well-housed government officials, who however, are not necessarily very committed to living in the town. Many of the migrants, and in particular the recently transferred government officials, appear to regard the town mainly as a place to work for a limited period of time, and not so much as one in which to live permanently. This is clear from the fact that so many of these migrants maintain a house elsewhere, while a considerable group even has wife and children staying 'at home', rather than joining them in the town. Housing conditions also did not provide strong evidence for social integration. Such differences may be explained by spatial and social-structural time-lags in the dimension of social modernization between major and minor centres, which itself of course is strongly related to differences in economic modernization.

There are also differences in migration characteristics as a spatial reflection of both economic and social change. The migration to both types of towns has become increasingly permanent in nature over the years. This is evident from their high population growth rates and their changing population structures. It was found that in Manzini the age and sex structure of the population had changed towards the 'normal' national structure, so that by 1982 it differed only slightly from it. Perhaps more important is the finding that only a small proportion of the urban households consisted of one person (i.e. the typical migrant-laborer household), while the great majority consisted of (complete) families. This was taken as evidence for the increased permanency of rural-urban migration. The same was reflected in the motivations of the migrants which showed, that a very large proportion of the migrants had come as dependents. Of further relevance is the fact that a large proportion of the urban migrants did not maintain connections with their original home area and therefore, apparently were firmly established as urbanites.

Migration to growing small centres did also show signs of becoming increasingly permanent in nature, but not quite to the extent that this has occurred in the large towns. This is evident again from the changes which have occurred in their population structures. In the regional centre of Nhlangano it was found that the age and sex structure of the population had changed from a very unbalanced one in 1976 to a more balanced one in 1984, but not quite as much as had happened in Manzini. Also, compared to Manzini, there was a larger proportion of migrants directly motivated by employment and a correspondingly smaller proportion who had come as dependents. Similarly, the Nhlangano population had a larger element of one-person households and fewer family households, while more migrants in this town maintained connections with their rural homes than in Manzini.

These facts appear to justify the conclusion, that the change from labour migration to permanent rural-urban migration, while noticeable with respect to both large and small towns, has progressed further in the former than in the latter. This appears to indicate that although such a transition follows a regular pattern, the spatial progression of it is uneven.

A lower degree of social modernization in small towns seems to be responsible for this and this in turn may be related to factors such as the temporary nature of the stationing of government officials in such towns, the physical and psychological proximity of the rural home for the large group of migrants from the hinterland region, which allows for frequent contacts and for families to be based there.

The implication is that policies to stimulate the growth of small towns must take into account not only the economic consideration of job creation and attracting companies, but also the social dimension of life in the town. The government as a big employer might play an important role in this by promoting the employment of local inhabitants in small centres in preference to transferring people from the large towns; and further through the promotion of the effectiveness of local government by decentralising functions and funds to the regional capitals. In the social sphere, a reduction in the frequency of transfers, the promotion of home ownership (rather than providing subsidized housing) for government employees and the expansion of family-housing schemes might also contribute to more stability of the urban population and its further growth. Where government employment is important the salaries of the employees provide a good basis for the growth of service functions; these should be stimulated to become fully regional in scope by, for example, improvements in public transport from the surrounding region.

But such measures by themselves are not sufficient to effectuate an increase in the proportion of the national population living in small towns. In order to achieve that, enormous growth of existing centres would be necessary (e.g. in Swaziland about 10 percent per annum in each centre). To expect such growth is unrealistic and what thus is needed for the government is to plan for additional small but viable centres to be established. In Swaziland there are company towns with deliberately restricted development strategies. Some of these have sufficient growth potential and should be stimulated (e.g. by uplifting their legal status to that of a town) to develop into small towns with a variety of roles to play in regional perspective. There are other very small centres, which with proper planning of e.g. regional roads, schools and hospitals (today often planned without reference to such potential centres) could be stimulated to develop into small regional service centres.

CHAPTER SIX
Small Towns, Labour Markets and Migrant Absorption: an Example from Northern Costa Rica

Arie Romein

This study deals with the absorption of migrants in the labour markets of small and intermediate urban centres. It places this theme into a regional perspective. To date, still little is known about this way of functioning of the smaller urban centres in Third World countries: little research has been undertaken on both the structure of labour markets in these centres and the absorption of migrants with due attention to the characteristics of their regional contexts.

Most studies on Third World urban labour markets and migration in the past decades were limited to large cities, especially national capitals, and hardly placed these topics into a regional perspective. With regard to smaller urban centres, broadly defined as centres up to 100,000 inhabitants, academic interest has focused more strongly on their role in regional development, and has emphasised their productive functions (trade, services, and processing of regional agricultural products) rather then their absorption capacity for migrants. The discussion has been confined mainly to their *possible* role in stemming the flow of migrants from their own rural hinterlands to primate cities. This role, however, has hardly been examined explicitly nor placed within the context of characteristics and dynamics of the regional hinterlands.

Our case study is one of the first attempts to analyse and understand the absorption of migrants in the labour market of a lower order urban centre within a specific regional context in Latin America. It presents an example from the Costa Rican town of Ciudad Quesada. Cd. Quesada is the administrative capital and largest urban centre (approximately 20,000 inhabitants in 1990) of the region Huetar Norte, an agricultural colonisation area in the northern lowlands of Costa Rica (figure 6.1).

The Regional Setting of the Study

In the late 1920s, Huetar Norte was still one of Costa Rica's geographically most isolated and least populated areas. Colonisation of the area had been only incidental and most often temporary. This peripheral character of the region started to change only after the 1929 Wall Street Crash. The international economic crisis in the wake of the Crash caused dramatically decreasing employment and income opportunities in the Costa Rica's two leading export industries coffee and bananas. Growing rural unemployment and landlessness in these industries' core areas generated a growing volume of migration to Huetar Norte and the colonisation of the region turned into a permanent, large-scale phenomenon.

Within the region, deforestation and soil exploitation due to this increased influx of colonists caused erosion and decreasing soil fertility. As a consequence, much newly occupied land was again abandoned after some years and rural colonisation frontiers set in motion from the foothills in the region's southern fringe into the lowland zone to the north. Although the 'frontier economy' embraced an ever increasing area, it remained geared to subsistence production for decades as the proper conditions for commercial agriculture hardly existed throughout the first half of this century.

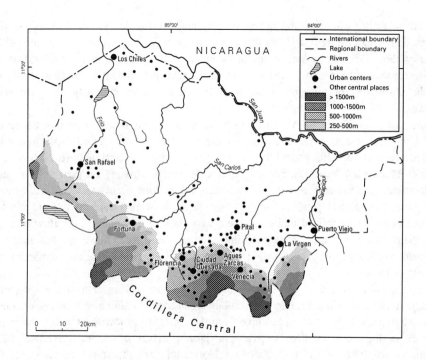

Figure 6.1: Huetar Norte

The 1940s were a period of increasing social instability and political violence in Costa Rica. This ended up, in 1948, in a six week civil war. After the war, a new reformist movement took over state power from the coffee oligarchy and set in motion far-reaching economic and social reforms under the header of the new development strategy *desarrollismo*. Where the agricultural sector was concerned, *desarrollismo* was after modernisation into a capitalist and highly productive industry of large and medium-sized farms. An important objective in this respect was the expansion and diversification of export-oriented production. To realise these reforms, state intervention in agricultural expanded enormously.

In Huetar Norte, the modernisation, commercialisation, expansion and diversification of agriculture accelerated a process of integration of the region into the national economy. Most subsistence cultivation was substituted by the production of food crops for the regional and domestic markets, export crops like coffee and sugar cane, and livestock. Many new colonists were attracted to the region by these improved opportunities for commercial agriculture and the frontier moved further. Moreover, the increasing commercialisation of agriculture was accompanied with new relations of economic dependency between large *hacendados* and small producers, leading to the latters' indebtedness and loss of land. For a growing number of these new landless, moving on and clearing a new plot of land further north was the only alternative (Manshard 1974). Where permitted by the road network, they mixed subsistence production with some commercial crops.

The development of Huetar Norte entered a new phase when Costa Rican planners predestined the region to become one of the country's leading export beef producers in the early 1960s. Demand for beef was high in North America and large areas of unused land in the region could still be converted into pasture. In order to assist the take-off of beef export industry, the state invested large sums in the necessary infrastructure and allocated huge amounts of credit to private ranchers. This policy was quite successful: export-oriented cattle breeding expanded impressively. The value of total beef export grew from 3 million United States dollars in 1962 to 28 million in 1972 and to 71 in 1980 (Fernandez 1988, p. 73; UNO 1985, p. 231). Clearly, this success would not have been possible without World Bank loans. The Bank made promotion of cattle breeding one of the cornerstones of its policy in Central America.

The rapid expansion of the beef sector in Huetar Norte has been accompanied with further polarisation of the region's land tenure structure. The World Bank and the Costa Rican authorities strongly agreed that large ranches would produce much more efficiently than small ones (Keene 1980) and credits were mainly supplied to very few large ranchers. These started not only to acquire vast quantities of state reserve land, usually virgin forests, for conversion into pasture, but also to buy out smallholders. Besides a growing concentration of land ownership in a few hands, cattle breeding in Huetar Norte has also been characterised by a very extensive use of land. One explaining factor is the prevalent type of credit, having been more proper for the acquisition of land and cattle than for the improvement of pastures and the animal stock (Aguilar & Solis 1988). Due to its extensive use of land, cattle breeding in Huetar Norte has provided relatively little employment. The estimated annual labour input per hectare in the region at the end of the 1970s was, for instance, 178 man days for coffee, 76 for sugar cane, 49 for rice and only six for cattle breeding (IFAM 1976, p. 20).

Due to these characteristics of the rapidly expanding cattle sector, the rural employment situation in Huetar Norte started to deteriorate in the 1960s. More and more peasants and agricultural labourers were evicted and forced to search for alternative means of subsistence. Colonisation of new land remained the traditional escape valve. Data from the last three Costa Rican population censuses (DGEC 1965, 1976, 1987) make it very enticing to assume an accelerated process of expulsion of rural labour from the southern part of Huetar Norte contributing to the development of active colonisation frontiers in its extended northern lowlands since the 1960s. This traditional way to cope with the loss of land and employment became supplemented with *precarismo,* the illegal occupation of either state or private land

and migration to urban centres. Although the majority of the bought-out peasant and unemployed agricultural workers may have migrated to the Greater Metropolitan Area (GMA) around the national capital San José, Costa Rica's largest urbanised region, some others settled in smaller urban centres within Huetar Norte.

The 1970s in Costa Rica were characterised by seemingly uncontrolled government spendings and rapidly rising public and balance of trade deficits. Instead of introducing an austerity programme to cope with these deficits, the political elite increased foreign borrowing. At the end of the decade, the country's debt service capacity further deteriorated because of the second oil crisis, increasing interest rates on loans, and deteriorating terms of trade for its agricultural exports. Consequently, the per capita foreign debt of Costa Rica skyrocketed to one of the highest in the world. In the beginning of the 1980s, the debt burden became uncontrollable and the country plunged into a severe socio-economic crisis.

The new Monge administration (1982-1986) responded to the crisis by putting sharp limits to imports and cutting back subsidies on items like fuel and public services. Parallel, it applied to the IMF, World Bank and USAID for new loans. These institutions were willing to provide large sums, but attached strict conditions in order to curtail the strong state intervention and regulation of the economy that had, in their view, caused the crisis by disturbing the functioning of the free market in the preceding decades. Under the header of structural adjustment, they strongly demanded more cutbacks in public expenditures, further expansion of agricultural export production, and the privatisation and liberalisation of the economy. The Costa Rican political spectrum, in which a lobby against state intervention and public spending had increased in strength during the previous years, was receptive for such demands and neo-liberalism has become the ruling economic ideology since the second half of the 1980s.

In flat contradiction to the structural adjustment theory (and propaganda), state intervention in agricultural production and marketing has not been reduced since the mid 1980s. Instead, it shifted unilaterally to export-oriented agriculture, in particular to new, non-traditional products like ornamental plants, citrus fruits and pineapple. These products have been heavily promoted by increasing regulations and every time more costly subsidies and other financial incentives. Because the package of incentives appears to have created a myriad of advantages for large capitalist investors over small local farmers to break into the exports of *no tradicionales,* export-oriented agriculture in Costa Rica has become increasingly dominated by large monocrop plantations. Many of these plantations are owned by transnational corporations. The production and marketing of the basic grains maize, beans and rice for the domestic market on the other hand, have been subjected to increasing liberalisation. The protective wall against competitive imports and the guaranteed forestalling and highly subsidised minimum prices that this segment enjoyed during the previous decades were cut back or eliminated. Consequently, most basic grain producers suffered a deteriorating income position and many went under. In addition to the deteriorated income position of small farmers, the salaries of agricultural wage labourers have also been put under pressure. To improve Costa Rica's competitiveness for *no tradicionales* in the foreign markets, a policy to lower their production costs has been introduced (Garnier 1989). The effect on salaries has been particularly serious because the IMF, World Bank and USAID have promoted the same few crops in all structurally adjusting countries in Central America and the Caribbean. Hence, the neo-liberal development strategy since the mid 1980s has had a downward effect on rural wages in Huetar Norte. In addition, the tendency towards growing inequality in the distribution of land ownership has not come to an end. The new plantation firms have purchased much of their large land stock from basic grain producers that had gone under due to a deteriorating income position.

To conclude, the agricultural production structure of Huetar Norte has been increasingly incorporated into the national and international space economy for the decades since the early 1950s. This process has been accompanied with extensive cattle ranching becoming the dominant type of land use and with a deteriorating access to land, employment and income for a growing stratum of the rural population. Both developments affected mostly the northern part of the region. Since the mid 1960s, a 'cattle frontier' has moved north to areas of large reserves of cheap land where the wet, badly drained soils are unsuitable for many crops (Skoruppa 1982, p. 279). Consequently, the agrarian structure in this area is strongly dominated by large, extensive ranches with very little wage labour (Altenburg et al. 1990). In addition to the rapidly increasing number of hectares under pasture, most *precaristas* settled in the north and the last active rural colonisation frontiers still absorbing evicted peasants are located there.

In the south of the region, the production of slaughter cattle is much less exclusive. Relatively more land is in use for the production of basic grains, dairy farming and traditional and non-traditional export crops. Due to the selective application of structural adjustment policies to the agricultural sector, the area under basic grains has decreased while the area put under *no tradicionales* has expanded rapidly since the mid 1980s. The newly established United States-Costa Rican joint venture *Tico Frut*, operating large citrus plantations and a juice factory, estimated that 17,000 hectares would be planted in the region at the end of 1992. Hence, the agricultural structure in the region's south is more diversified than in its north: extensive cattle ranches exist side by side with large foreignowned plantations, medium-sized modern farms, and small peasant farms.

In the wake of the moving colonisation frontiers over an increasing area of Huetar Norte, a spatial system of settlements has emerged in the region since the second quarter of the century. More recently, the rapidly increasing commercial agricultural production for both regional, domestic and foreign markets have induced the establishment of many new urban functions in these settlements. These new functions have been directly linked to both the production, marketing and transport of crops and cattle and to growing rural household spendings. In addition, the numbers of urban functions in the region's towns have accelerated because of an explicit regional policy that was formulated in the two National Development Plans of 1974-1978 and 1978-1982. To realise a reduction of the large socio-economic inequalities between Costa Rica's core region, the Central Valley in the middle of the country, and its peripheral regions, planners attempted to force national public and semi-public institutions with headquarters in San José to establish branch offices in newly designed planning regions. Huetar Norte is one of these regions. Overall, the region's present hierarchy of settlements is composed of three levels. The lowest level contains rural villages, the large majority of its settlements. The intermediate level is made up of a much smaller number of settlements that have developed into local service centres for their immediate rural vicinities. At the top of the hierarchy, Cd. Quesada (13,066 inh.) has developed into the region's leading service centre. Located in the 'bridgehead' of the earliest moving colonisation frontiers and today's most densely populated part of Huetar Norte, Cd. Quesada is both the nodal point of the region's transport network and its main gateway to the Central Valley. In addition, it has gained the status of administrative capital of the planning region Huetar Norte.

Due to both its situation and its administrative status, considerable economic interaction has developed between Cd. Quesada and its regional hinterland. Beside the broad range of petty retail shops, homebound service shops, and small repair and artisan's workshops that operate strictly local, the production structure of Cd. Quesada also includes larger, modern and well equipped shops, offices and manufacturing plants that maintain various types of economic linkages with the region. The town is first and foremost a central place that provides services and goods to final consumers and enterprises, both agricultural and non-agricultural, in Huetar Norte. Besides a large private service sector, many national public and semi-public institutions have established branch offices in Cd. Quesada, in particular within the framework of the regional development policy of 1974-1982. Overall, the area of influence of the service apparatus of Cd. Quesada covers the entire region while the smaller service centres of the regional hierarchy of settlements fill in local gaps in service provision (Nuhn 1989). The trading sector in Cd. Quesada includes agencies of the national beer company, national food chains, and some foreign soft drink marks which obtain their inputs from the GMA and distribute these all over Huetar Norte. Compared with its large trading and service sectors, the town's modern manufacturing sector is small. It is limited to a few medium-sized plants that process regional agricultural inputs into bulk quantities of milk powder, sugar, semi-roasted coffee, and timber. Apart of timber, most of these semi-manufactured products are transported to the GMA for further processing or to the nation's air and sea ports for export. In fact, the largest share of the agricultural production of Huetar Norte is not industrially processed in Cd. Quesada but in the region's countryside or in the GMA. Cattle even completely bypasses the industrial sector of Cd. Quesada. Hence, the importance of the regional hinterland to Cd. Quesada's production structure is first of all this area's role of consuming market for the town's trade and service sectors. Its importance as supplier of inputs of agricultural raw materials for industrial processing in the town or for trans-shipment trade to external locations is much smaller.

Having been Huetar Norte's largest and most articulated urban labour market already for decades, Cd. Quesada has developed into a major destination for migrants in the region. Migration to the town has mainly been undertaken by entire families for reasons of either finding better employment and higher incomes or transfer from offices in other locations, usually the GMA, of one (or a few) breadwinners. Concerning the size of migration, a household survey in the town in 1989-1990 (see note 1 for some remarks on the data collection for this study) revealed that the share of migrants in its population of fifteen years and over amounted to about one third for some decades. This share was 31.6 percent in 1973 and 32.7 percent in 1989.

Part of the recent inflow of migrants in Cd. Quesada consists of former peasants and agricultural workers from the region itself that have been evicted by the expansion of extensive cattle ranching and, more recently, monocrop plantations. However, this intra-regional migration from rural sites is only a minor share of total migration to the town. This has been due to two forces. First, many displaced peasants and agricultural workers have bypassed Cd. Quesada in favour of larger urban centres outside the region, but more likely in favour of the colonisation frontiers in its northern borderlands. As yet, not all these frontiers have been closed and migration to Cd. Quesada appears to have replaced migration to the frontiers only to a limited extent. Second, migration from the region's rural areas have been overshadowed by the inflow of migrants from the GMA with its population many times larger than Cd. Quesada itself. The regional offices of many public and semi-public institutions that have been established in the town have been staffed with many transferred employees from the GMA headquarters.

The Labour Market of Cd. Quesada

The analysis of the absorption of migrants in the labour market of Cd. Quesada is based on a segmentation of this market into employment sectors. This segmentation was supposed to reveal qualitative differences within the town's labour market. Much literature of Third World urban labour markets distinguishes employment sectors on the basis of characteristics of jobs. Often cited are certainty and regularity of jobs, employment status or labour relation, average length of working-week, and size of income. A quality of employment index on the basis of such characteristics was considered a necessary element of a classification of employment sectors in Cd. Quesada. In itself, however, such an index lacks sufficient analytical precision to distinguish and adequately describe employment sectors. Some other variables that are closely related to qualitative differences between jobs also need to be taken into account. The first one is the segment of production: major differences in many qualitative respects occur between jobs in what are often called the formal and the informal sector. Moreover, jobs also differ qualitatively between occupations within these sectors. The formal sector may for example include factory workers besides medical specialists. Therefore, occupation is another variable that needs to be included in the analysis of employment sectors.

To start with the latter variable, the various occupations practised by the working population in Cd. Quesada were classified into occupational groups. The following five were distinguished: white-collar occupations, sales occupations, blue-collar industrial occupations, non-white collar service occupations and agricultural occupations. Next, the division of the production structure of Cd. Quesada into segments started with a typology of surveyed enterprises and institutions[1] on the basis of differences in their level of technology, organisation of production, scale of operation, and economic and technical ways of conducting business. These four variables were each measured by one or more equally weighted and dichotomised indicators (Annex 6.1). On the basis of a low (0-3) versus high (4-7) sum of scores on the total set of seven indicators, a L(ow) and a H(igh) segment were distinguished. The average L-segment enterprise is less well-equipped, characterised by a less advanced organisation of production and a lower monthly turnover or budget, and makes less use of insurances, detailed bookkeeping, formal credit and separate bank accounts than the average H-segment enterprise or institution. After labeling enterprises and institutions L or H, branches of industry were grouped into a L or H segment. A branch was classified into either the L or H segment if a minimum of three-quarters of the surveyed enterprises or institutions had either a low or high sum of scores. For branches that were too heterogeneous to meet this criterion, a third, M(ixed) segment was designed in addition to the L and H segments.

Annex 6.2 presents the distribution of establishments in the various branches of industry over the H, M and L-segments of the production structure of Cd. Quesada. The four distinguished types of industry were found to be represented differently in the segments: the service and trade sectors are very well represented in the H-segment while the largest part of the industrial sector and even the entire repair sector are in the L-segment. This appears strongly related to the major function of Cd. Quesada for its regional hinterland, the leading central place that provides services and goods to final consumers and enterprises: the establishments in the H-segment maintain much wider networks of economic hinterland linkages than those in the L-segment.

The crosstable of occupational groups and production segments was used as the basis for the division of the labour market of Cd. Quesada into a limited number of employment sectors. Considering the agricultural sector a distinct production segment and leaving the unemployed out of consideration, five sectors were distinguished (Table 6.1).

Table 6.1 Employment sectors in the labour market of Cd. Quesada, 1990

	H	M	L	A	U
White-collar occupations	V	III	II	I	-
Sales occupations	IV	III	II	-	-
Blue-collar industrial occupations	IV	III	II	-	-
Non-white collar service occupations	IV	III	II	-	-
Agricultural occupations	--	---	--	I	-
Unemployed	--	---	--	-	0

A: Agricultural sector
U: Unemployed
Source: Household survey

The five employment sectors are described below. In addition to characteristics of the respective production segments and occupational groups, attention is also paid to the quality of employment. To measure this quality, three characteristics of jobs were selected: employment status, certainty, and part-time or full-time (Annex 6.3). These characteristics were considered dichotomised variables. The sum of scores on these variables were combined into an index that measures the quality of employment. The value of this index ranges between 0 (low quality) and 3 (high quality). Finally, the employment sectors are described by two characteristics of the working people. In general, the accessibility of specific occupations appears to be related to two personal characteristics of applicants; gender and level of education. Hence, the distribution of the economically active population over the employment sectors is associated with these personal characteristics. For the analysis, three levels of education were distinguished: low (has finished primary education at most), intermediate (has started secondary education, finished or not, or has started a vocational training without having finished it) and high (has finished vocational training or has started university education, whether finished or not).

The highest employment sector (V) in the labour market of Cd. Quesada consists of white-collar occupations in the H segment. It includes just over one quarter of the town's working population (Table 6.2). The large majority work in the service industry, in particular in public and semi-public institutions. These institutions operate in branches like public administration, law and order, insurance, public utilities, agricultural services and banking. Agricultural services are provided for example by the Ministry of Agriculture and Livestock, the semi-public institute CNP providing technical assistance and storage to producers of food grains, and private chambers of cattle and sugar cane export producers. The banking sector, moreover, supplies most of its credits to livestock breeding and export crop production. Most of the remainder of the workers in this employment sector were found in trade, in particular retail shops in non-daily goods and luxuries and wholesale businesses in foodstuffs and drinks. The make-up of the highest employment sector together with its size, reflect the role of Cd. Quesada as the main service and trade centre for the extended regional hinterland of Huetar Norte. Almost all workers in this highest employment sector practise medium to highly qualified professions like manager, medical doctor, teacher, or book-keeper. Hence, this employment sector is hardly accessible to applicants with only a low level of formal education. Gender, on the contrary, is not a barrier of entry at all, albeit some clear male and female occupations can be distinguished. Due to both the required higher levels of education and the public or semi-public status of the institutions offering the bulk of employment in sector V, the quality of work is rather high: only one-tenth of the workers do not enjoy a full time, permanent, and salaried job.

Table 6.2: Working population of fifteen years and over in Cd. Quesada by employment sector (%), 1990

V	IV	III	II	I	Total (n)
25.4	17.6	17.3	33,6	6.1	393

Source: Household survey

The next highest employment sector (IV) includes all other occupations in the H-segment of Cd. Quesada's production structure. One-sixth of all workers are to be found in this sector. Compared with the highest employment sector, workers in sector IV are less concentrated in public and semi-public service institutions; about one half of them were found outside these institutions, in either large retail shops, wholesale businesses and agro-industrial plants.

The workers in this employment sector practise non-white collar service occupations such as driver or guards, sales professions and blue-collar industrial occupations. The required level of formal education in this employment sector is considerably below that in sector V; the highest level is rarely demanded while additional specific skills are often obtained on the workfloor. In addition, opportunities for women to find a job in this employment sector are much more restricted than in sector V. Employment in this employment sector is clearly dominated by men, except the sales occupations. On the other hand, the quality of employment in this employment sector is hardly below that in sector V, despite workers' lower level of education. This is explained by the fact that the enterprises and institutions in this employment sector, as in V, belong to the production structure's H-segment in which almost all jobs are permanent, full-time, and salaried.

At the other extreme of the labour market of Cd. Quesada, employment sector II includes all employment in the L-segment of its production structure. It is the largest employment sectors in the town's labour market, containing one-third of the workers. In striking contrast with the two highest employment sectors, more than half of the workers in this one practise blue-collar occupations in small to very small and minimally equipped artisan or repair workshops. The major artisanal branches are tailoring and the production of wooden and leather articles like furniture, boots and horse saddles. The repair branch, entirely classified in the L-segment, includes a broad variety of workshops, specialised in items ranging from tree-trunk lorries to watches. A lesser, although still a substantial share of the blue collar workers in this employment sector earns a living in a great many petty retail shops, selling items like foodstuffs and second-hand clothes, or in homebound personal services like hair dressing. The only white-collar workers in this sector are the relatively highly educated free professionals in technical and economic services, such as lawyers, notaries, bookkeepers and accountants.

Because employment in the artisan and repair workshops is nearly exclusively carried out by men, they make up the large majority of the workers in this sector. For women, entry to this employment sector appears very difficult. They were only found in trade occupations and in some non-white collar service occupations such as domestic servant or in small beauty salons. The average required level of education is still somewhat below that in sector IV. In general, this means that the participation of female workers in the labour market of Cd. Quesada relatively increases in employment sectors where the required level of education is higher, a rather unusual finding. The quality of employment in sector II is quite low according to local standards; only half of the workers enjoy a job of highest quality, i.e. score 3 on the index.

The fourth employment sector (III) is an intermediate one between the two highest ones V and IV and the low one II. The proportion of workers found in this employment sector is equal to that in IV, one-sixth. It comprises all employment in the M-segment of the production structure. This segment is a mixture of small-scale, poorly equipped trades and medium to large-scale businesses in catering and retail. Manufacturing is represented only by a few saw mills and the repair sector is not represented at all. Due to the large importance of trade and retail branches in the M-segment, almost all workers in this employment sector practise sales occupations and non-white collar service occupations. The intermediate position of this sector is accompanied with both an intermediate average quality of employment and an intermediate required average level of education. The level of education is slightly over that in sector II but much below that in sector V. As sectors IV and II, male workers outnumber female workers, albeit to a lesser extent. Neither of the two most important branches in this sector III, catering and retail, exclude female workers.

The final employment sector (I) comprises employment in the agricultural sector in Cd. Quesada's rural hinterland. It is the smallest of all five employment sectors with only six percent of the economically active population. The required level of education is very low. With the exception of a very few agronomists, workers need no more than a basic level of literacy. When it comes to the sex of workers, it is the only employment sector where no females were found at all. Finally, the quality of employment in this employment sector is the lowest of all employment sectors.

The Absorption of Migrants Into the Labour Market

Absorption of workers into labour markets has a quantitative and a qualitative dimension. As to the former, Table 6.3 shows that the share of employed migrants is only slightly below the share of employed locals. Moreover, this difference is compensated by a slightly lower unemployment rate among the migrants. Although migrants may be less well informed about the town's labour market at their time of arrival than locals, a chisquare test (at the 0.05 level) reveals that the participation rates of migrants and locals in the labour market of Cd. Quesada do not differ significantly.

Table 6.3: Activity status of locals and migrants in Cd. Quesada (%), 1990

	Locals	Migrants	Total
Employed	54.8	50.4	52.7
Unemployed	5.4	3.5	4.6
Housewife	23.5	36.3	28.1
Studying	12.6	1.0	8.3
Disabled/aged	3.7	8.8	5.5
Total (n)	488	284	776

Source: Household Survey

As to the qualitative dimension of labour absorption, the distribution of migrants versus locals over the employment sectors in the labour market of Cd. Quesada suggests some striking differences (Table 6.4).

Migrants are relatively overrepresented in employment sector V of white-collar occupations in the public and semi-public institutions, whereas locals are overrepresented in sector IV, comprising sales occupations in wholesale businesses, supermarkets, and specialised retail shops; blue-collar occupations in agro-processing plants; and non-white collar service occupations in modern, private service institutions.

Table 6.4 Working locals and migrants by employment sector in Cd. Quesada (%), 1990

	V	IV	III	II	I	Total (n)
Locals	22,4	20.9	16.5	35.0	5.1	254
Migrants	30.9	11.5	18.7	30.9	7.9	139
All	25.4	17.6	17.3	33.6	6.1	
Total (n)	100	69	68	132	24	393

Source: Household survey

Taken together, these two employment sectors include 43 percent of the working *quesadenos*. Among the other 57 percent, active in either sector III, II or I, the representation of locals and migrants is fairly evenly balanced. This explains why a chi-square test revealed that the distribution of the two sub-populations is not significantly different across the employment sectors, despite the over and underrepresentations of migrants and locals in the employment sectors V and VI respectively.

Overall, it can be concluded that the absorption of locals and migrants in the labour market of Cd. Quesada differs significantly neither in a quantitative nor in a qualitative respect. This indicates that the employed migrants are far from a homogeneous group in the sense that they would be very much concentrated in one, or probably two related employment sectors. This leads us to assume that the absorption of migrants is a differentiated process, related to the existence of some clearly distinguishable sub-groups of migrants. Some studies of small urban centres in Latin America suggest that there is a relationship between the geographic characteristics of the places of origin of migrants and their opportunities in these centres' labour markets. Migrants from urban origins are presumed to have better employment opportunities than migrants from rural origins due to higher levels of education and skills. To examine this supposed relationship in Cd. Quesada, its migrant population will be divided into sub-groups according to the urban or rural nature of the regions of origin.

The working migrants in Cd. Quesada could be divided into three sub-groups by geographical origin. The first sub-group originates from the GMA. This region offers the best educational facilities at both the secondary and tertiary levels of the country. It is supposed that this sub-group has the highest average level of education. For this reason of a high educational level, the few Costa Ricans who returned home after having attended specialised courses in foreign countries were included in this sub-group. Assuming that this sub-group has also the widest range of urban skills - the GMA has Costa Rica's largest and most diversified urban labour market and provides most high skilled jobs they may have obtained jobs in, on average, higher employment sectors than other migrants in Cd. Quesada.

The second sub-group of migrants origin from the rural areas, villages and local service centres in Huetar Norte. The educational apparatus in these locations is of a much lower order than in the GMA. The only educational facilities at a higher than primary level are a few institutions for agricultural vocational training. In addition, the variety of highskilled urban sector jobs is considerably less in these locations in Cd. Quesada's regional hinterland. In fact, a considerable proportion of this area's economically active population depend on work in the agricultural sector hence, this sub-group of intra-regional migrants is the lowest educated one and has most probably obtained only a few higher urban skills.

The third and last sub-group of migrants origins from 'other regions' than the GMA and Huetar Norte. It was supplemented with the few migrants from foreign origin, mostly the other Central American republics, with nationalities other than Costa Rican. These other regions include all types of settlements except the metropolitan one, i.e. rural sites, local service centres, small towns and medium-sized urban centres. The single most important 'other region', the western Central Valley, is located closely to the GMA with its large, diversified urban labour market and extended education apparatus. Migrants from this region, therefore, may have commuted to work or school in the GMA before they moved to Huetar Norte. Hence, the proportion of migrants from these other regions' rural areas or local service centres that have received tertiary education and have obtained some type of higher urban skills may be above that of the intra-regional migrants.

Based on their supposed differences in level of education and skills, it may be assumed that the three sub-groups are absorbed in different employment sectors in Cd. Quesada's labour market. Competition for jobs in specific employment sectors may be assumed to be very limited in particular between migrants from the GMA and from Huetar Norte. Table 6.5 shows that the proportion of employed are smallest in the sub-group of migrants from the rural areas and local service centres in Huetar Norte and largest in subgroup of migrants from the GMA.

Distinguishing the two categories of employed versus non-employed migrants, a chisquare test nevertheless does not allow us to conclude that their participation rates differ according to their region of origin. Hence, in a strictly quantitative sense, the various sub-groups of migrants have not been absorbed to a different extent into the labour market of Cd. Quesada.

Table 6.5 Activity status of sub-groups of migrarts according to their geographical origin in Cd. Quesada (%), 1990

	GMA	Huetar Norte	Other regions
Employed	57.3	47.1	47.3
Unemployed	4.9	5.9	0.9
Housewife	30.5	40.0	37.0
Studying	2.4	--	0.9
Disabled/aged	3.7	7.1	12.9
Other	1.2	--	0.9
Total (n)	82	85	108

Source: Household survey

To examine the distribution of the working migrants in the various sub-groups over the employment sectors, low cell frequencies forced us to join some of the sectors in order to apply the chi-square test. These were first II and I, employment in the small-scale and poorly equipped shops and workshops in the L-segment and in agriculture in the rural hinterland. Employment in both sectors is mainly non-white collar, of relatively low quality, dominated by men, and requires the least formal education. Next, the sectors IV and III, non-white collar occupations in the H-segment and all types of occupations in the M-segment, were combined. In terms of the required level of education, opportunities for women, and quality of employment, this combination of IV and III falls comfortably between sector V and the combined sectors II and I.

Table 6.6: Sub-group of migrants by employment sector in Cd. Quesada (%), 1990

	GMA	Huetar Norte	Other Regions
V	46.8	23.1	22.9
IV + III	27.7	23.1	43.8
II + I	25.5	53.8	33.3
Total (n)	47	39	48

Source: Household survey

Table 6.6 shows that the sub-group of migrants in Cd. Quesada with a metropolitan background is best represented in sector V, white-collar employment in the H-segment of the town's production structure. On the other hand, the sub-group of migrants from the rural areas and local service centres in Huetar Norte is predominant in the sectors of what is almost exclusively other than white-collar employment in the small scale repair and artisanal workshops, personal service shops, and agriculture. Finally, the sub-group of migrants from the other regions of Costa Rica is most strongly represented in the intermediate sectors of other than white-collar employment in agro-industrial plants, wholesale and modern retail businesses, and in the catering sector. The chi-square test revealed that this distribution of sub-groups of migrants from different geographical origin across the (combined) employment sectors in Cd. Quesada is statistically significant at the 0.05 level. Hence, in a general sense absorption in Cd. Quesada is indeed a differentiated process in which migrants with a more urban background find employment in higher sectors.

A chi-square test corroborated the above supposed dependence of sub-groups of migrants' levels of education on their geographical origin. This finding arises the surmise that it is the level of formal education of working migrants rather than their place of origin as such that determines their over and underrepresentation in certain employment sectors in the labour market of Cd. Quesada. A three dimensional log-linear test confirms this surmise, although it has to be taken into account that this conclusion is based on data that had to be 'compressed' for reasons of a technical prerequisite for application of the test. The intermediate and high levels of education were combined into one category that was contrasted with the low level and the labour market was re-subdivided into the employment sectors IV and V (the H and M-segments) versus I, II and III (the L-segment and the agricultural sector).

Table 6.7: Sub-groups of working migrants in Cd. Quesada by sex (%), 1990

	GMA	Huetar Norte	Other regions
Male	70.8	65.0	68.6
Female	29.2	35.0	31.4
Total (n)	47	40	51

Source: Household survey

A second personal characteristic of working migrants that may determine their over- or underrepresentation in certain employment sectors in the labour market is gender. Table 6.7 shows that the labour participation rate of males is considerably higher than that of females among migrants in Cd. Quesada: in all sub-groups about two-thirds of the employed are male. Due to this similarity, a chi-square test revealed no statistically significant difference in sex rate of the workers between the sub-groups. Therefore, the employed migrants from rural villages and small local service centres in Huetar Norte for example, are overrepresented in the low-quality, non-white collar types of employment in the L-segment as well as in agriculture, because they have the lowest levels of education and not because males constitute a much larger majority in this sub-group than in others. Thus, contrary to their level of education, gender is not an important determinant of the working migrants' over or underrepresentation in specific employment sectors in the labour market of Cd. Quesada.

To conclude this section, it may be observed that the process of differential migrant absorption in the various sectors of employment in the labour market of Cd. Quesada shows a correlation with the size and geographical range of the external economic linkages maintained by the corresponding enterprises and institutions. Well-educated migrants from the GMA were clearly overrepresented in the employment sector that corresponds with the public and semi-public institutions and the medium-sized to large modern enterprises that maintain important backward linkages with the GMA and forward linkages with the entire regional hinterland. The production segment that depends almost exclusively on economic linkages at the local level of Cd. Quesada and its immediate vicinity on the other hand, is largely the domain of lowly educated migrants from Huetar Norte.

Summary and Conclusion

Our study of Cd. Quesada in the agricultural colonisation region of Huetar Norte in Costa Rica aimed at contributing to the understanding of the absorption of migrants into labour markets of lower order urban centres. To this end, the study first related the functioning of Cd. Quesada to the development of its regional hinterland. The town appeared as the region's leading central place as it satisfies in the first place the demand for urban goods and services by the regional population, i.e. either as entrepreneurs or consumers. In addition, the town has a manufacturing sector that processes regional agricultural raw materials into semi--manufactured goods. This sector however, is relatively small. The functioning of the town is not limited to this regional level. On the wider, extra regional level, Cd. Quesada is the intermediate node in chains of distributive and/or collecting trade between Huetar Norte and the metropolitan area of San José and in the downward chain of the provision of services from this metropolitan area to the regional hinterland in Huetar Norte. In addition to the functioning of the town on two different geographical levels, it is also bypassed by flows of goods and services between its rural hinterland and the metropolitan area. The functioning of the town in the wider world has resulted in a specific structure of its labour market. It was sub-divided into four employment sectors on the basis of differences in the production segment of enterprises or institutions, the occupational structure of workers, the quality of employment, and required personal characteristics of workers. These four employment sectors are supplemented with a small one in agriculture outside the town. The town's highest employment sector contains highly educated white collar workers in highly educated occupations in modern public and semi-public service institutions, trade enterprises and agro-industrial manufacturing plants. Quality of work is high and gender is no barrier of entry to applicants. The majority of workers in the lowest sector on the other hand practise blue collar jobs in what may be called the informal sector of petty shops and workshops. The quality of employment is relatively low. but education is no barrier of entry.

The absorption of migrants in the labour market of Cd. Quesada was found to be a differentiated process. Different sub-groups of migrants have been absorbed in different employment sectors. Broadly speaking, the best educated working migrants in the town originate from larger urban centres and are overrepresented in the employment sector of high quality white collar jobs in the town's production segment of public and semi-public institutions and modern trade and manufacturing enterprises. At the other extreme, the least educated migrants originate from rural areas and local centres in Huetar Norte itself and are predominant in lower quality jobs in small workshops and petty trades.

The example of Cd. Quesada makes obvious that the process of absorption of migrants into the labour markets of lower order urban centres indeed needs to be analysed within their specific regional contexts. The size and nature of the various employment sectors in the town, with different accessibility for different sub-groups of migrants, are related to its regional context. Moreover, the systems of lower order urban centres within their regional contexts should not be conceived as closed systems. Both regional towns and rural hinterlands may function within a wider geographical framework and, in addition, against the background of government development policies, both sectoral and regional. A large variety of economic linkages and policy relations with the extra regional world may affect the specific process of migrant absorption.

Note

1. The bulk of the primary data for this study was collected by means of two surveys in Cd. Quesada in 1989-90, one among households and one among enterprises and institutions. The household survey aimed at obtaining data on employment, personal characteristics of workers and migration to Cd. Quesada. The objective of the second survey was twofold: a typology of enterprises and institutions according and an examination of the geographical patterns of their forward and backward economic linkages.

The predominant type of household in the northern Costa Rican cultural context is the nuclear family, living in its own, separate house. The house, then, was selected as the unit of sampling in the household survey. A detailed map, made by the Costa Rican statistical office DGEC for the 1984 census showed individual houses. After updating in the field, it was employed as a sampling frame. The first step in the sampling of households was a division of the town in types of neighbourhoods. On the basis of field observation, five types were distinguished: the commercial centre; parcelled-out, older neighbourhoods around this centre; new, spontaneous and not-parcelled out lower-income neighbourhoods at the town's outskirts; a large, mixed-income neighbourhood (San Martin) in the south-western part of the town, and *urbanizaciones,* middle and high income neighbourhoods, some still partly under construction, at various outskirts of the town. A second step was to take a sample of houses from neighbourhoods by type. This was done by means of two-stage sampling. In all neighbourhoods, one in three so-called *segmentos,* small spatial units of around 60 houses indicated on the DGEC census map, were selected randomly and from each of these sampled segments approximately one in four houses was drawn. The resulting grand sample included 283 households. The overall sampling fraction was about 1: 11.

The sampling frame of the survey of enterprises and institutions was also based on the updated DGEC map. Walking through the streets of Cd. Quesada, all visible enterprises and institutions were marked on these maps. The total number thus mapped amounted to 802. Although the completeness of the sampling frames constructed in this manner was not guaranteed - some small-scale activities carried out by individuals or on family basis may be excluded -, no better sampling frame was available. One possible alternative, the municipality's list of establishments revealed more imperfections. Some economic branches in Cd. Quesada were represented by a very small number of establishments. Unfortunately, these are in particular branches that deserve our special interest because of supposed more intensive and more diversified economic linkages with the rural hinterland. To avoid a (near) absence in the sample of such branches, a stratified sample was taken instead of a simple random one. The first stratum included three of the above mentioned branches, the agro-processing plants, business and agricultural services, and retail of farm inputs, whereas the second stratum comprised all other economic branches. The strata included 69 and 141 establishments and their sampling fractions were 0.51 and 0.21 respectively.

CHAPTER SEVEN
Reception Centre and Point of Departure: Migration To and From Nuevo Casas Grandes, Chihuahua, Mexico

Tine Béneker & Otto Verkoren

The literature about rural or regional development in Latin America, frequently states that small and intermediate towns attract and absorb (fair numbers of) rural migrants. Precisely because of this absorptive capacity, quite a few students, planners and policy-makers consider Latin American small and intermediate towns instrumental in the stemming of the flow of migrants to larger cities and metropolitan areas (see e.g. Borsdorf, 1981; Mathur, 1982; Sibbing, 1984; Hardoy and Satterthwaite, 1986; Hansen, 1990).

Remarkably though, we have little empirical evidence to substantiate the absorptive capacities of the small and intermediate towns. Although Latin American literature on internal migration movements is surprisingly large, most scholars focused their attention on migration to the *large cities* (see e.g. Muñoz et al. 1982; Billsborrow et. al., 1984, 1987; Cantu & Gonzalez, 1990; Preston 1996; De Oliveira & Roberts, 1996). Indeed, migration studies about Latin America's large cities are roughly available from the early sixties onwards, and they cover almost all Latin American countries, as well as almost all important cities. In comparison, the number of migration studies dealing which deal with small and intermediate cities in Latin America, is much smaller (nice examples are e.g. Verduzco, 1984; Evans, 1989; Van Lindert & Verkoren 1991; Velasquez Gutierrez & Arroyo Alejandre, 1992; Romein, 1995). Next, only few of these studies were undertaken before the eighties, as a consequence of which they cover only a limited timespan. And finally, the small town studies cover, geographically, a limited number of countries and regions In fact, with the information currently available, it is far from easy to corroborate or to refute existing ideas on the role of smaller cities in the stemming of the flow to the larger urbs.

This contribution tries to shed some light on this topic, by focussing on migration flows to and from the town of Nuevo Casas Grandes (NCG), a proud city located in Mexico's Great North. To that end, the second paragraph casts a quick glance in a few salients aspects of the State of Chihuahua, the regional background of this study. Next, in paragraph three, the main dimensions of migration to NCG will be analysed, using data from a grass-roots-level survey carried out in the city. Paragraph four gives an idea of out-migration from NCG, by tapping another section of the same database. In the last paragraph of this contribution the findings of the NCG-study will be put in perspective, through a comparison with a number of other recent studies.

The Setting

With almost 250,000 square kilometres, the State of Chihuahua is the largest of the Federal Republic of Mexico, making up for some 13 percent of national territory. In demographic terms, Chihuahua does not lead the pack. Since the State boasts only some 2,5 million inhabitants, it occupies the 10th place among the Mexican Federal Entities, well behind the population-rich ones such as Mexico, Distrito Federal, Vera Cruz, Jalisco or Puebla.

It logically follows that the State of Chihuahua is a thinly populated territory. In fact, the average population density of the state barely reached 10 pp/km2 in 1990. Such an overall figure is quite misleading however, because Chihuahua's population is very unevenly distributed. Very large tracts of the State's territory (such as the Chihuahuan desert, or the higher parts of the Sierra Madre Occidental) are virtually unsettled. In fact, more than 85 percent of Chihuahua's population lives on less than 40 percent of the State's territory! Still, the average population density of this "permanently settled area" only reaches a meagre 18,3 pp/km^2.

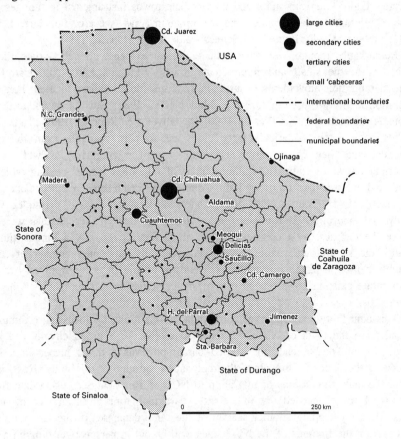

Figure 7.1: The settlement system of the State of Chihuahua

A closer look at the population distribution reveals even a much stronger concentration. The rural population is overwhelmingly concentrated in a limited number of ecological niches (such as the fertile river valleys, or the intermont basins). In these ecological niches, population densities easily surpass 100 pp/km2, sometimes even 150 pp/km2. However, the majority of Chihuahua's population lives in cities (in Mexico "urban" relates to centres with 2.500 inhabitants or more). Already in the early sixties, the State's level of urbanization surpassed 50 percent. Since then, the urbanization process went on with a surprising pace. The urban population skyrocketed and by 1990, at the time of Mexico's XI Census of Population and Housing, more than three quarters of the state's population was living in urban areas. In the process, Chihuahua's rural population diminished. Indeed, since the sixties, more and more rural regions turned into clear-cut expulsion areas, many of which began to show absolute population declines since the eighties.

Chihuahua's settlement structure is highly unbalanced. The urban hierarchy is headed by two large cities (Ciudad Juarez and Ciudad Chihuahua), both amply surpassing the 500,000-mark. Next, the hierarchy shows three secondary cities (i.e. Delicias, Hidalgo del Parral and Cuauhtemoc), with populations close to 100,000 inhabitants. The third tier of the urban hierarchy (the size-bracket of 10,000 - 50,000 inhabitants) consists of nine cities. Some of these tertiary cities have relative small spheres of influence, since they are very much part of the 'market area' of a nearby larger city. Others, such as NCG or Ojinaga, still command a vast rural area 'of their own'.

The fourth tier of the urban hierarchy consists of a fairly large number of small urbs. Although only few of them surpass a number of 5,000 inhabitants, most of them certainly are of local importance, if only because they generally are the *cabeceras* of the *municipios* (i.e. the administrative centers of the municipalities). The lowest tier of the settlement hierarchy consists of the smallest localities. As a rule of thumb, these settlements are tiny, widely scattered and overwhelmingly rural. Indeed, some 95 percent of the 11,000 localities of the State of Chihuahua, does not even reach 500 inhabitants!

The growth rates of Chihuahua's urban settlements vary enormously, largely due to differences in migration rates. On the lowest level of the urban hierarchy, the population growth figures are very modest, frequently lower than the average population growth of the State (which in the period 1980-90 was 1.8 percent). Here, out-migration is simply outstripping natural population growth, a phenomenon which also holds for quite a few of the smallest towns, the *cabeceras*. Apparently, Chihuahua's smallest settlements are (no longer) able to retain their youngsters, a phenomenon which clearly comes to the fore in their ageing population pyramids. In contrast to the smaller settlements, the tertiary cities (like NCG), show relatively high growth rates, normally well above the average population growth of the State. Here, in-migration plays a conspicuous role, frequently already for several decades. Although these tertiary centers have carved out their own migration fields, a careful analysis of census data since the sixties suggests, that these centers are also part of the migration fields of the larger cities. Hence, in-migration (from the rural areas) goes hand in hand with out-migration to the larger urban centers. The same phenomenon seems to be the case with the secondary cities in Chihuahua's urban system. While they draw 'their migrants' from a relatively large migration field, they are clearly subjected to the attractivity of Cd. Juarez and Cd. Chihuahua. Consequently, Chihuahua's urban system can be described as a nested hierarchy, with clearly overlapping migration-fields.

The Changing Production Structure of Nuevo Casas Grandes

The town of Nuevo Casas Grandes is located in the Northwestern section of the state of Chihuahua in a flat basin, wedged between the foothills of the Sierra Madre Occidental. The city commands the sparsely populated area of Chihuahua's Northwest, which covers some 35.000 sq.km^2 and which embraces eleven municipios.

Originally, NCG was created in 1892 as a railroad station with a few shacks. But already by the beginning of this century, it had turned into a thriving little town. Well before the second World War, NCG's local industries processed regional agricultural products, such as food, fruit, meat, and milk. Its wood manufacturing sector rose to regional fame in the processing of timber, construction materials and furniture. A fair number of repair shops in the city specialized in servicing mechanized agriculture. And last but not least, a wide array of services successfully catered both for the urban and rural populations. After the Second World War, NCG's production structure was further diversified, when various new plants were opened which produced fabricated metal products, while the agricultural repair sector became involved in the manufacturing and assembly of (agricultural) machinery. In fact, as time went by, NCG overpowered the other small centres in the Northwest, a phenomenon which was backed by the *carrefour* location of the town.

In 1960 the small town boasted some 16,000 inhabitants. But in 1970 this number was almost doubled, a phenomenon which was, of course, strongly influenced by in-migration. In the next decade, NCG's economic growth slowed down somewhat, as a consequence of the slump in the (agricultural) business cycle. Consequently, the cities' attractivity for regional, rural migrants was eroded a little bit, while quite a few citizens out-migrated. At the end of the seventies, NCG's number of inhabitants had 'only' grown with some 7,000 persons, a growth rate of 1,4 percent per annum, well under the rate of natural population of that period. However, the eighties brought a new impulse for NCG's production structure. Small wonder that in-migration in the city picked up again. Consequently, NCG's annual rate of population growth shot upwards to almost 5 percent per year, more than twice as high as the annual natural population growth rate. In 1990 NCG had already 49,154 inhabitants, a figure which in the mean time might very well have increased to more than 60,000!

Although the town's economic base had rapidly broadened since World War II, agriculture continued to be the mainstay of the production structure for quite some time. As Table 7.1a shows, even in 1970 the agricultural sector employed more than a third of the labor force. Since then, the winds of change have altered the town's production structure profoundly. First, a number of foreign assembly-plants (the so-called maquiladoras) arrived, in the metal and machinery sector, which boosted low-skilled employment in town, especially for young females. Next, the agricultural processing activities expanded impressively. Together, these two types of production were a tremendous impulse for the manufacturing sector. Furthermore, a rapid expansion of NCG's service activities took place, in the public as well in the private sphere.

Table 7.1: The changing labor force of the Municipio of Nuevo Casas Grandes, Estado de Chihuahua, 1970-1990

a) Industry and the labor force

Industry	1970		1990	
	Abs.	Percent	Abs.	Percent
Agriculture	2605	36.4	2739	17.7
Manufacturing	969	13.5	3886	25.1
Construction	263	3.7	1319	8.5
Transportation	357	5.0	766	4.9
Commerce	916	12.8	2143	13.8
Other Services	1372	19.2	3808	24.6
Others/Unspecified	676	9.4	837	5.4
Total	7158	100.0	15498	100.0

b) Occupational Groups and the labor force

Occupational Group	1970		1990	
	Abs.	Percent	Abs.	Percent
Professionals/Technicians	383	5.4	1047	6.8
Managers/Administrators	205	2.9	645	4.2
Office Workers	566	7.9	1329	8.6
Traders/Sellers	555	7.8	1684	10.9
Crafts-, Repairmen & Factory Workers	1465	20.4	5534	35.6
Agricultural Laborers	2454	34.3	2543	16.4
Other Service Workers	1224	17.1	2105	13.6
Others/Unspecified	306	4.2	611	3.9
Total	7158	100.0	15498	100.0

Source: Dirección General de Estadistica, IX Censo General de Población. 1970. Estado de Chihuahua, Mexico D.F., 1971; INEGI, Chihuahua, Resultados Definitivos Tabulados Basicos, Tomo II, XI Censo General de Población y Vivienda, 1990. Aguascalientes, 1991

Here, sectors like education, health, banking and insurance, retail and wholesale increased in importance, the construction sector thrived, while the repair and maintenance sectors also profited.

Table 7.1a clearly shows the dramatic change of NCG's labor force: in absolute terms the number of workers in Manufacturing quadrupled, as a consequence of which this sector became the largest one, absorbing a quarter of the total labor force. Also, the tertiary sector underwent a sizeable increase, from some 37 percent in 1970 to 43 percent in 1990.

As Table 7.1b shows, the sectoral changes between 1970 and 1990 had far reaching consequences for occupational structure. On the one hand, the so-called White-collar segment of the labor force (i.e. the Professionals/Technicians, the Managers/Administrators and the Office Workers) increased fairly. Among the Blue-Collar Workers (here defined as the Crafts- & Repairmen & Factory Workers together with the Agricultural laborers) a clear-cut shift towards non-agricultural employment is discernable.

Migration to Nuevo Casas Grandes

The migration survey in NCG related to some 300 households, randomly selected through a city-wide sample, whereby 427 migrants were encountered (Stam, 1991).[1] Where did these migrants come from? An overwhelming majority originated from the State of Chihuahua itself. Indeed, more than 85 percent of the migrants were Chihuahuenses, while the rest came mainly from the neighboring states of Sonora, Sinaloa and Durango. As Figure 7.2 shows, short distance migration predominates. Indeed, the bulk of NCG's migrants came from areas relatively nearby. Apart from a few migrants from the municipio of Nuevo Casas Grandes itself, almost one third moved in from immediately adjacent municipios. Another third came from the municipios in the *Sierra Madre Occidental*, the mountain region immediately south of Chihuahua's Northwest.

Since almost all of these municipios are rural (Madera being the one exception), one would presume that regional migration to NCG is basicaly a rural-urban affair. In this respect some caution is warranted however. With the Mexican distinction between "rural" and "urban" in mind, only some 60 percent of the incoming regional migrants effectively came from rural areas. The rest of them originated from urban *cabeceras*. Hence, while settling in NCG is a rural-urban move for the majority of the regional migrants, the city also receives quite some regional migrants who typically experience a move up the urban settlement ladder. This bi-modal split is clearly reflected in the work-experiences of both migrant groups, as well in their educational attainment levels. While the rural migrants had almost exclusively agricultural backgrounds, the ones from the *cabeceras* generally worked in construction, trade, transport, repair work etc. Next, the migrants form the small towns were slightly better educated. While many rural migrants had not completed primary education upon arrival in NCG, the majority of the migrants from the cabeceras at least possessed a certificate of primary education.

Apart from the regional migrants, NCG accommodates also a completely different group of people. Some 30 percent of the migrants in NCG originated from the larger urban areas, such as Cd. Chihuahua, Cd. Juarez, Cauhtemoc, Delicias or Hidalgo del Parral), or even from large cities from outside the state, such as Zacatecas, Durango or San Luis Potosi. In comparison to the low-skilled regional migrants, these urban migrants had higher educational attainment levels, generally at least completed secondary education. Indeed, the survey clearly showed that it was the urban-origin migration which brought clerical workers, administrators, (middle-) managers and other professionals to town. As such, this phenomenon was intimately associated with the diversification of the town's economic base. In the sixties, when NCG was still a modest regional service center, this type of migration was still quite insignificant. In fact, the survey detected in this respect only a few school teachers, a tax officer, as well as some policemen and nurses. From the mid-seventies however, when NGC's manufacturing and service sectors began to grow, demand for skilled labour rose steeply. Since neither the local labour market, nor the rural surroundings were able to meet this demand, a rapidly growing in-migration of urban-trained people was a logical result.

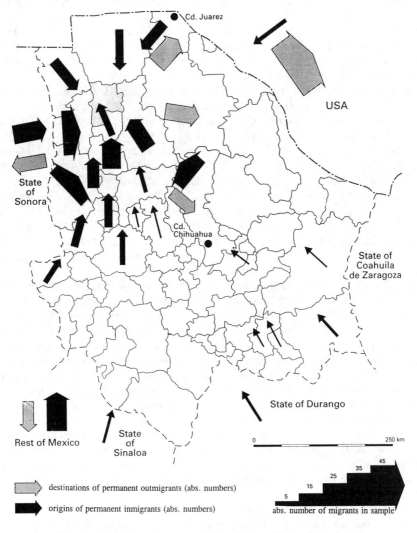

Figure 7.2: Migration to and from Nuevo Casas Grandes

Migration to NCG is (still?) very much a 'family affair'. Indeed, an overwhelming majority of the migrants (82 percent) arrived in NCG as part of a (nuclear) family. Most of the arriving families (i.e. some 75 percent) are relatively young, i.e. in the child-bearing or child-rearing stage. Their arrival was normally the result of a direct, one-step, migration from a location in NCG's hinterland. The rest of the incoming families experienced much more complicated movement-patterns, with three, four of five steps, including large cities in Chihuahua or elsewhere.

Of course, one also finds singles among the movers to NCG, both males and females. However, this group is relatively small: it makes up for some 15 per cent of the sampled migrants. Young singles clearly dominate in this group, more especially young females, who mainly are in the 15-19 years age-bracket. As such, this phenomenon can be easily related to growing labour demand in the maquiladora-sector, which is especially keen on employing young (and relatively little skilled) females. Generally these young women come from villages in NCG's immediate hinterland and they live in town within the families of relatives. The male single movers are a highly diverse group, among which the so-called returnees from the USA stand out.

Why do people migrate to NCG? For 50 percent of the interviewed persons the motive to settle down in NCG was clearly economically inspired. The majority of the regional migrants did not have a pre-arranged job on arrival. On the contrary, most of them hoped that they would get one soon, through the backing of relatives or friends already present in town. And most of them got a job quickly: more than 90 percent of the rural migrants was employed within one week after arrival, be it frequently in short-term contract work. However, the analysis of the employment histories showed, that NCG's labour market clearly offered possibilities for occupational mobility. Within a year after their arrival, more than 60 percent of the regional migrants had effectively left their original buffer-function (with its very low wage and limited job-security), and changed it for a better paid job with a fixed position or a long-term contract. Most of the urban migrants however, did possess a job-contract on arrival; indeed, they knew when and where to start and under which conditions.

Among the non-economic motives to come to NCG, two phenomena stand out. About 20 percent of the migrants came to NCG for *'family-reasons'*. In this category we find e.g. the young women who settle down with the spouse they just married, or the elderly persons who joined the household of one of their children. On the other hand we also find here the family-reunions which are intimately related to, what has become known as, *delayed family migration*. Basically this means, that the head of the household initially moves alone and "sends" for his/her family to come over, when an income opportunity and/or a dwelling has been secured. For some families, this form of family reunion takes only a few weeks, for others it takes several months or more. The other important non-economic reason to move to NCG relates to the town's educational facilities, an item which was mentioned by some 15 percent of the interviewed heads of household. Both in NCG's hinterland, as well as in the Sierra region, the number and the variety of schools is low and their quality is frequently perceived to be inadequate. For quite a few rural households, the movement to NCG was even specifically related to the parental desire of better educational possibilities for their offspring.

What about the regional migrants who bypass NCG and go to other places? The analysis of Mexico's censuses since the 1970's suggest, that a great many Northwestern Chihuahuans migrate from their villages and small towns without settling down in NCG. In fact, it is possible to estimate from the census data that no more than 40-50 percent of the regional migrants actually goes to NCG. The rest opts for other possibilities elsewhere. Despite a relatively strong and diversified production structure, the town of NCG certainly is not *the* reception center for *the* regional migrants. Consequently, in discussing the role of NCG in stemming *the* flow to the large cities one has to be careful.

The available census data are not revealing much about the socio-economic characteristics of the regional migrants who bypass NCG. However, it is possible to come up with some clues, by using the information from a survey that was carried out in Cd Juarez.[2] As to the educational attainment levels, it was clear that the settlers from the Northwest in Cd. Juarez were better off, than the ones who went from the same region to NCG. One has to bear in mind however, that the educational backgrounds of these two groups are not fully comparable. Many of the heads of households (and their wives) who came to NCG from the Northwest in the seventies and eighties simply did not have that much educational opportunities in the countryside. Many villages and hamlets did not have a school; and if one was present, ti was frequently so crowded, that it had been decided to run a morning and an afternoon "shift". Especially in the eighties, the Mexican government put a lot of effort in improving the educational situation in the countryside. Hence, many youngsters who left in the late eighties and early nineties had really had the opportunity to get (primary) education.

Comparing the work-experiences of the regional migrants who went to NCG and to Cd. Juarez, it clearly showed that those who went to the large city had been less involved in agricultural jobs. Indeed, in this respect it was clear, that it was NCG which accomodated the farmers and the farmhands. Finally, the types of work which the leavers (eventually) performed in Cd. Juarez, could easily be classified as "better" (i.e. more remunerative, more stable, with a higher security) than the ones of those who went from the region to NCG.

Migration from Nuevo Casas Grandes

Now that we have an idea of in-migration in NCG, we can shift our attention to those who left town. Each of the sampled heads of households was asked a number of questions about members of the household, who departed from NCG in the year preceding the survey, for a period of more than three months. Through this approach, data were gathered about 129 persons (88 males and 41 females). Since their leaving, none of them had returned to settle again in NCG. In fact, only 37 persons had visited NCG again since their departure, although some 65 percent of them frequently sent letters, postal money orders, or cheques to their parents.

Where did the out-migrants go to? It would be only logical to presume a relatively limited outflow to the surrounding rural areas, *as well as* an important one to (Chihuahua's) large cities. The survey data showed that both of these expectations were correct (see also Fig. 2) As to the migration down the settlement hierarchy, the survey showed only eleven persons who had traded NCG for a smaller place in the State's Northwest; another five went to the smaller localities in the Sierra Madre Occidental. All of these out-migrants were middle-aged persons (i.e. over 35 years of age), and all were said to have returned to their *'tierra',* to their birthplace, either to retire or to take up an agricultural life again.

Furthermore, the survey indicated that 47 out-migrants (36,4 percent) departed for larger urban places in Mexico. Not surprisingly, Cd. Juarez and Cd. Chihuahua came to the fore as important destinations, but also some other lager cities, further away, like Tijuana and Mexicali in Baja California, and Metropolitan Monterrey. Remarkably, this group of outmigrants invariably consisted of young males and females in the age-classes of 15-19 yrs and 20-24 years. Almost 70 percent of them were said to have left NCG for employment reasons. For another 20 percent educational motives were stated (which were almost exclusively related to higher vocational training or a university degree). As to the rest of the out-migrants going to the larger cities in Mexico, a set of highly divergent motives were given, among which family-reasons (such as marriage or family reunion) was most conspicuous.

Unfortunately, the socio-economic information in the NCG-survey about the out-migrants had its drawbacks. Some of the data of the earlier mentioned Cd. Juarez survey enabled us to fill in a part of the puzzle. Apart from the 81 migrants who came to Juarez from NCG's hinterland, also 37 migrants from the town were 'detected' in the large city on the border, who had left the city between 1985 and 1993. With only a few exceptions, this group consisted of young singles, 26 males eleven females. They had moved directly to Cd. Juarez, most of them for employment - reasons, a few of the to study. Slightly more than half of them (i.e. 20) were born in NCG, hence could be considered as natives; the others did have NCG as last place of residence, but were born in other places, almost exclusively in Chihuahua's Northwest of the central parts of the Sierra Madre.

The educational attainment levels of those who departed for Cd. Juarez clearly differed from the migrants who came to NCG from its hinterland. Indeed, all of the Juarez-migrants had a primary school diploma and many of them had even enjoyed (a few years of) secondary education and/or vocational training. In fact, from an educational perspective the "leavers" were much more qualified than the 'rural' migrants who flocked into NCG.

About three-quarters of the female migrants from NCG stayed with family in Cd. Juarez and quickly found a job, invariably in one of the cities' maquiladoras. Some 25 percent followed a different pattern however, they became domestic servants, living in the house of their employers. Quite a few of these *Empleadas Domesticas con cama adentro* were even "recruited" for that jobin NCG, by their prospective employer! The male migrants preceeded their education in Cd. Juarez (35 percent), or quickly found a job, generally through a help of relatives of friends. More often than not they started in all kinds of odd jobs at the bottom of the employment scale, but generally they quickly moved up the occupational ladder to better-paid, and more or less fixed, jobs in construction, manufacturing or service-work.

The largest group of out-migrants from NCG (59 persons or 45,7 percent of the registered movers) went to the U.S.A., by and large illegally. Again, this group mainly consisted of youngsters, which were principally males. In fact, in this category the survey traced only 13 females! As such, migration to the U.S.A., legally or illegally, is a wellknown phenomenon (see e.g. Massey et al, 1987; Jones 1996; Zabin & Hughes, 1995; Durand et.al 1996). However, most available literature on the topic perceives it as a phenomenon, which was principally related to the Mexican rural environment. Seeing it also (and that importantly) popping up in a small Mexican town was a bit of a surprise. The major gateway to the US were Cd. Juarez, and the small towns of Palomas (Chihuahua) and Agua Prieta (Sonora), whereby most of the emigrants had, according to the NCG-informants, successfully relied on the services of a paid "broker" (*coyote*). Remarkably, the majority of the emigrants did not live and work in the US border zone. On the contrary, NCG's emigrant population was widely scattered over the USA (with clear concentrations in

Los Angeles, Phoenix, Albuquerque, Houston and Chicago). Neither did the emigrants, according to the informants in NCG, earned their money mainly in US agriculture, e.g. as cattlemen, tractor drivers, or day labourers. More often than not, they worked in (the bottom of) the service sector, e.g. as dishwashers, window cleaners, gardeners, guardsmen, or in the construction sector. And although most of them planned to become typical sojourners in the US, month after month their stay was lengthened, even up to the point that the regular money transfers to NCG began to run dry.

Discussion

In Latin American (planning-oriented) literature, small towns are frequently perceived as '...transmitters of development, little engines of growth, places for work and checks on the rural exodus' (Hinderink, 1997, p. 187). In this respect, the absorptive capacities of small towns for regional (rural) migrants have frequently been stressed as an instrument to stem the flow of migrants to the large(r) cities. How important is NCG in this respect? Let us try to make a few comparisons with relevant, recently done, other 'small town' migration studies.

As to the migration to NCG, it is clear that the town recruits the bulk of its migrants from its own hinterland. Thus, the pattern of migration to NCG is dominated by short distance moves. Next, most migrants who flock into town come, with their families, directly from farmsteads, hamlets and villages. Migration to NCG can thus be labelled as a one-step, rural-urban affair, which bears the hallmark of family-migration. In this respect, our findings corroborate to a large extent with earlier studies on migration to small towns (such as Adamchak, 1978; Hoenderdos & De Regt, 1991; Velasquez Gutierrez & Arroyo Alejandre, 1992; Béneker, 1997).

Apart from an important rural influx, NCG also receives fair numbers of urban migrants, who originate from higher-order centres. Again, this phenomenon links up with earlier studies (like e.g. De Regt, 1987; Harfst 1988; Woltman, 1990; Van der Luit 1992; Romein, 1995). Moreover, the NCG-study also suggests that the importance of this inter-urban migration down the settlement hierarchy has been increasing over time, as the demand for skilled employment (i.e. blue collar jobs as well as white collar ones) was growing. Apparently, under conditions of rapid economic expansion, the capabilities of the small town local labour markets fall short of demand, as a consequence of which the import of qualified personnel from elsewhere becomes necessary.

As to the household composition of the incoming migrant-flow, in-migration in NCG is firmly dominated by young families, which generally are in the childbearing and child rearing stages. For these families, the employment opportunities offered by the NCG-labour market plays an important role, like the possibilities for a better education of the children. In this respect, the NCG-study also dovetails nicely with earlier research (see e.g. Massey et. al., 1987; Du Toit, 1990; Root & De Jong, 1991; Bradshaw, 1995). For young, single migrants NCG's appeal is much more limited. That is probably why so many youngsters in Chihuahua's Northwest are bypassing the town, once they have decided to migrate. Nevertheless, the arrival of the maquiladoras, has certainly given a boost to NCG's attractivity for young, single females. But even the latter phenomenon has been found earlier (see Woltman, 1990; Hoenderdos & De Regt, 1991; Béneker, 1997).

The educational attainment levels of the in-migrants in NCG show striking differences.

The lowest educational qualifications clearly pertain to the regional, rural in-migrants. The in-migrants from the smaller urban places in the region were, in this respect, just a little bit better off. Educationally, the migrants who came form the higher order centres were the best equipped, as a rule of thumb even better than NCG's native population. It logically follows, that the three in-migrant groups clearly have different positions on NCG's labour market. Indeed, the rural in-migrants in NCG are strongly concentrated in the bottom of the employment scale, while those from the large cities are preponderant in the middle and top-end. In their studies, Kleijburg (1988), Béneker & Hooijmaijers (1990), Van Lindert & Verkoren (1991), Rodriguez & Smith (1994), Romein (1995) came up with comparable results.

Migration from NCG is, on the one hand, closely connected with inter-urban migration. Indeed, it strongly relates to youngsters, who move up the urban hierarchy, mostly for employment and/or for study-reasons. On the other hand, for a segment of NCG's young males, the possibilities in the USA, *al otro lado de la frontera,* also are very attractive. In this respect, NCG certainly is not an exception. On the contrary, as Castles & Miller (1993) have pointed out, the USA became a very important migration destination for Latin Americans. Although the rural and urban areas of Northern Mexico may participate a little stronger in this US-oriented migration, there is no doubt that Central and Southern Mexico are also heavily involved in this kind of spatial mobility, just like e.g. Central America (see Zabin & Hughes, 1995; Durand et. al. 1996; Béneker, 1997).

Looking back, the migration data gathered in NCG compare fairly well to the findings of other small town migration studies. Indeed, we feel that NCG, with respect to many migration-related phenomena, can be considered a nice, representative case. Our data show, that NCG effectively is a reception center for migrants from Chihuahua's Northwest. However, the town principally attracts a very particular slice of the Chihuahua's Northwestern section, viz. young families, the parents of which have relatively little education and whose job-experiences dovetail with the opportunities offered to them by NCG's labour market. Apparently, NCG's capacity to absorb regional migrants is quite selective: suitable for some, less for others. And in view of the bypass of the town from the Northwest, NCG's absorptive capacities are also quite limited. A probable majority of the migrants population of Chihuahua's Northwest, does not even seem to consider NCG as an option to settle in. The nature of migration to a small town simply differs a lot from the migration to large(r) cities!. Just because small towns are mainly attractive for a specific group of migrants from 'the countryside', one cannot perceive these centres as *the* option for *the* rural, regional migrants.

While the labour markets of small towns absorb a specific portion of regional migrants, they are apparently unable to retain all of their labour force. Many youngsters leave for other places, mostly the children of both natives and migrants. Our data clearly show in this respect that NCG is also a point of departure for another generation, who as a rule of thumb has a higher educational attainment level. Indeed, NCG is both an reception center and a clearing house, a phenomenon which was also encountered by e.g. De Regt & Hoenderdos (1991) or Béneker (1997).

In comparison with other small towns, the socio-economic situation of NCG may be labelled as relatively positive. After all, NCG is a typical regional service center, which commands a large rural area without serious competition from other centers (except Cd. Juarez, much farther away). Moreover, NCG is a thriving town, with a diversified production structure, rapid economic growth and quite sunny perspectives. Despite these advantages, NCG's impact on regional migrants is a modest one. Hence, less well endowed small towns, might even be expected to play an even smaller role as attraction center. The material presented in Béneker & Romein (1994) about the Costa Rican towns of San Isidro del General and Quesada clearly gives some food for thought in this matter.

Notes

1. Migrants were defined as persons born outside the city, who were at least 15 years old when they arrived in town. Consequently, the data do not related to most of the young children who arrived as family-members.
2. In the Cd. Juarez survey 597 households were interviewed, whereby information was gathered about 2828 persons, of which 1612 were migrants (using the same definition as in the NCG-survey). Among these migrants, 81 came from the ten, so-called rural municipalise of Northwest Chihuahua. Furthermore, the Juarez survey encountered 39 migrants from the municipality of NCG, of which 37 came from the town.
3. However, at the time of the survey in Cd. Juarez, almost 80 percent of them was married and had a family in the child-bearing or childbearing stage.

CHAPTER EIGHT
Town and Hinterland in Central Java, Indonesia

Milan Titus & Alet van der Wouden

Since the middle of the 1980s the role of small towns in regional and national development also gained increased attention in Indonesia (NUDS, 1985). In many developing countries, as well as in academic circles such towns are seen as instrumental in achieving developmental goals. However, empirical evidence on the present functioning of those small towns in developing countries, and in Indonesia in particular, is still rather scarce. This is why geographers of the University of Utrecht embarked upon a research programme on the functioning of small towns and regional development in the Serayu Valley in Central Java. The Gadjah Mada University in Yogyakarta acted as a counterpart in this programme. The research activities have focused on three towns, viz. Banjarnegara, Purbalingga and Wonosobo, each having the status of a district (*kabupaten*) capital. A more limited and supplementary survey has been carried out in a smaller sub-district centre i.c. Purworejo-Klampok.

The purpose of this chapter is to present the most relevant research findings with respect to the production structure and production relations of these small towns, their functional relationships with the rural hinterland and higher order centres (outside the region) and their employment structure and labour markets. These subjects are analysed through a structural approach explained in chapter three of this book.

Enterprises were chosen as the basic unit in the analysis of the production structure, whereas households were the basic units in the analysis of the employment structure. Data were collected by random sampling procedure on the basis of enterprise counts in selected urban areas, and on the basis of a complete list of households in each town.

The Research Area

The research area (see figure 8.1) is the Serayu Valley region, an accidented and densely populated river basin in the southern part of Central Java, with some 3,2 million inhabitants and an average density of 902 per km^2 (1980). The population is mainly concentrated in the lower central and western parts of the region, although most of the mountain areas are densely populated as well. The region's peripheral position is reflected in its remoteness from the major thorough fares and economic centres of Java, the absence of big cities (over 500.000 inhabitants), and in a higher than average percentage of the population depending on agriculture (more than 80 percent).

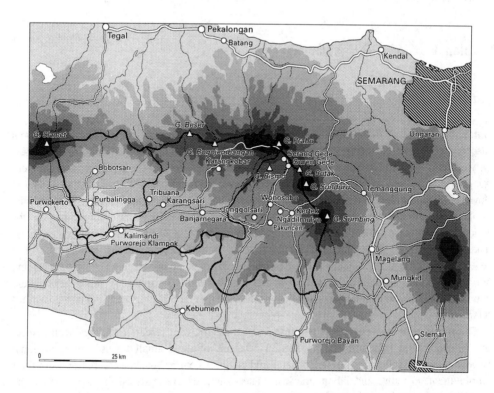

Figure 8.1: The Upper Serayu Valley Region

Yet, intensive and commercialized types of peasant-agriculture have developed here both in the western and central lower areas (mainly wet-rice and sugar-cane cultivation) as well as in the higher mountain areas to the north and east (market-gardening of vegetables and tobacco). A more subsistence-oriented type of agriculture with rainfed food-crops (cassava and maize) is found in the hilly uplands south and north of the central valley. Because of the high population pressure, landlessness has become a major problem, affecting some 30-40 percent of the rural population in the lowland areas, and fragmentation of land has reduced the average farm-size to less than 0,5 ha. Consequently, most small farmers and landless people have to find incomes in both farm- and non-farm activities. None of the main urban centres in the Serayu Valley displays a very dynamic growth pattern as is shown in table 8.1.

Table 8.1: Population growth of the main urban centres in the Serayu Valley Region, 1930-1980

	1930	1961	1930-1961 in percent	1971	1961-1971 in percent	1980	1971-1980 in percent
Purwokerto	33,266	80,556	4.6%	94,023	1.7%	149,782	6.5%
Banjarnegara	---	13,351	---	16,008	2.0%	19,355	2.3%
Purbalingga	16,435	22,698	2.3%	24,434	0.8%	29,766	2.4%
Wonosobo *)	10,701	16,170	1.7%	18,757	1.6%	26,058	4.3%

*) Latest figure based on different administrative borders from previous years
Source: De Jong and Ligthart 1986, p. 30

With the exception of Purwokerto, the dominant centre of the region, and recently Wonosobo due to administrative border changes, the three research towns have experienced growth rates not very much in excess of the natural growth of Java ($\pm$ 2 percent). This slow growth indicates their little attractiveness to migrants from rural areas and other urban centres. Of the three small towns, Wonosobo with its hinterland of highly commercialized mountain crops appeared to be the most dynamic one, whereas Banjarnegara with its less developed and less accessible hinterland was found to be the most stagnant town. Their strong service-oriented economic structure is reflected in their employment structures (see table 8.7).

Considering the heavy dependency of these towns on their administrative and commercial functions, it seems logical that these two pillars of the urban economy are closely interrelated. Sometimes even to the extent that government officials are indulging in both activities at a time. Government expenditures for public services and regional development programmes therefore, play a crucial role in the structure and functioning of these centres, as has been demonstrated in the 1970s and early 1980s when the oil-boom fuelled Indonesian development efforts. In the next section we shall discuss the structural model that may help to explain these complex interrelations and policy-effects on the urban economic structure and its functioning.

It was assumed that because of its peripheral position the small towns in the Serayu Valley would be relatively undisturbed by external interventions from higher-order centres outside the region. On the other hand, recent interventions from government programmes aimed at increasing agricultural productivity, improving infrastructural facilities and expanding the institutional framework, may have affected the economic structure and role of these service centres. This situation enables us to study the dynamics of structural change at the local (intra-urban) and regional (inter-urban and rural-urban) levels.

The Urban Production Structure and Hinterland Relations of the Three Small Towns

In fact Geertz (1965) was the first and only author to analyze the production relations in such type of urban centres in his well known study on a small town in East Java. In this study he opposed a bazar-type economy to a firm-type economy, more or less along the lines of the then well established theory of economic dualism.

As this dualistic model has been severely criticized for its rigidity i.e. its denience of the existence of functional relations between the two sectors and its lack of understanding for the dynamism and efficiency of certain bazar type or informal sector activities, there was clearly a need for further explorations into this field from a structuralist point of view (see chapter 3). With the exception of Forbes' study in Ujung Pandang (Forbes 1981), however, no such effort has been undertaken for other towns in Indonesia. Moreover, Forbes' study has focused only on the petty commodity mode of production, so that an overall picture of the urban economic structure and its functioning was still lacking.

Towards a Structural Model of the Small Town Economy
Facing the problem of getting a hold on the complex structure of the small-town economy in Indonesia, Titus, de Jong en van Steenbergen (1986) have tried to develop an operational model for analyzing its production structure and functional relationships. It is this model - briefly discussed in chapter three of this book- that will be elaborated and tested within the context of our research area.

The model starts from the basic assumption that the unequal process of capitalist penetration and transformation leads to a multitude of enterprise-types rather than to their rigid dichotomization into mutually excluding sectors. On the other hand however, the model should be flexible enough to provide for a sectoral division 'a posteriori' if higher levels of analyses require such generalizations, e.g. when studying intersectoral and rural-urban relations at the local and regional levels. Of special relevance in this respect is the distinction between petty-commodity and corporate enterprise production in order to analyze processes of conservation and dissolution.

The classification of enterprise-types first of all is based on so-called mode of production characteristics which are structuring the urban economy and hence are called structural characteristics. Specific indicators of the capitalist mode of production are e.g. production for a market, private property of means of production, the use of wage labour, spatial and functional separation of production and household activities, intensive use of capital and formal sources of credit, and last but not least, the raising of profits for capital accumulation and reinvestment. The score of an enterprise on this scale of indicators determines its capitalist or non-capitalist nature. It should be noticed however, that not all of these indicators are equally relevant in a Third World urban economy. Production for the market for example, is a useless indicator in an urban economy where self-sufficiency already has become impossible. In this respect even the smallest petty-producers have one or two capitalist characteristics as has been adequately demonstrated by Van Dijk (1982) in Salatiga. Another problem rises with the indicator based on the separation of production and household activities. Whereas this criterion applies very well to manufacturing activities or public services, it does not for petty-traders or independent transport and construction workers, because here the physical separation of work and household is inevitable but without meaning for the mode of production as such.

Next to these structuring characteristics our model is based on so-called form-characteristics, discriminating enterprises according to their petty-commodity or corporate character. These characteristics primarily apply to scale of enterprise, registration status, the use of modern technology and management-techniques (e.g. book-keeping). Although these form-characteristics are to some extent related to the structuring characteristics, the connections between them are mostly of an indirect nature and certainly not *'per se'*. Finally, the scores on both types of characteristics have been taken as entries to a matrix-system in which each enterprise or activity finds its own position. The following example shows how this works out for the enterprise survey in Wonosobo, when these enterprises are classified according to seven structuring and five form-characteristics:

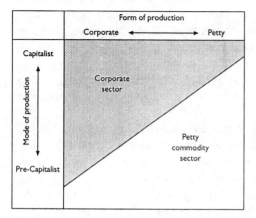

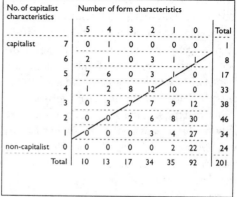

Source: De Jong & Ligthart 1986, p.18.

Figure 8.2a: A model of the urban production structure according to enterprise characteristics

Figure 8.2b: Distribution of enterprises in Wonosobo according to number of structuring characteristics as listed in appendix

A specification of these types of characteristics and the percentage of enterprises involved in each of them is shown in annex 8.1. As expected, the form-characteristics seem to have a stronger effect on discriminating between petty and corporate types of enterprise than the structuring characteristics.

As mentioned before, the matrix should also provide for the possibility of identifying enterprises according to their predominant petty or corporate enterprise character. This sectoral division *'a posteriori'* however, is not without ambiguities. First, because there seems to be a continuum of enterprise-types instead of a dichotomous distribution and secondly, because it is not clear whether this sectoral division should be made according to form or structuring characteristics. If petty-commodity production is only regarded as a form of production, the problem might be solved quite easily, for example by drawing the line at two or three form-characteristics while taking care of the fact that petty producers should at least have smallness of scale as their main feature.

Most authors however, agree on the fact that petty- or corporate types of production also involve differences in modes of production (cf. Gerry 1978, Forbes 1981). This assumption is corroborated by our enterprise data from Wonosobo, which suggest that there is some correlation between form-characteristics and structuring characteristics; a majority of the enterprises operating in a non-capitalist way indeed seems to be of the petty-commodity type (see bottom-right end of the matrix), whereas a larger part of the enterprises operating in the fully capitalist way is of the corporate type. Thus the dividing line between the sectors does not run vertically or horizontally, but diagonally (see figure 8.2).

Comparing the percentage of petty and corporate enterprises scoring on structuring and form characteristics respectively (annex 8.1), it shows that the differences between these sectors are more outspoken with the form characteristics than with the structuring characteristics. This seems to reflect the fact that capitalist transformation already has pervaded all types of petty activities without changing their scale of production, technology or management, so that their petty features more or less are being conserved. Consequently, form characteristics have a greater weight in the sectoral division between petty and corporate enterprises than structuring characteristics.

The enterprise matrix of Wonosobo clearly shows the overwhelming predominance of petty-producer activities in the small-town's production structure. This is completely in line with Moser's observation that the small-town economy is characterized by informal enterprises of which an overwhelming majority is engaged in tertiary activities, employing only one worker, i.e. the self-employed (Moser 1984). Opinions however, diverge on the main reasons for this predominance of petty or informal producers in small towns. In a later article Davila & Satterthwaite (1987) suggest that many informal enterprises exist only because there is an insufficient demand to support formal sector enterprises. This in contrast with the situation in larger urban centres where informal enterprises exist because they are functional to the formal sector. Therefore, the authors expect the informal sector to diminish in importance as soon as a small town becomes more prosperous and less isolated from the national economy. Our enterprise-matrix data from Banjarnegara, Wonosobo and Purworejo-Klampok however, do not show significant differences in the share of petty-commodity production (80-85%) despite considerable differences in economic dynamism and size. Probably because their roles as regional service centres are not sufficiently differentiated. Another possibility however, is that the growth of formal or corporate sector enterprises entails a parallel growth of the informal or petty commodity activities, the latter subsisting on the increasing circulation of money and goods generated by the former (cf. Titus, De Jong & Van Steenbergen 1986; Hinderink & Titus 1988).

Intra- and Intersectoral Relations at the Local Level
With the matrix set out above it is not only possible to identify types of enterprises according to their petty or corporate character, but also to study the relative position of the

various branches of activity (e.g. market-trade, shopping, manufacturing or transport) as well as their complex interrelations. Taking examples from the most important branches of economic activity in town, i.e. toko (shop)-trade and pasar (market)-trade, we may see that in both branches there is a variety of enterprise-types according to the structuring and form-characteristics, which makes it impossible to classify either of them as fully corporate or fully petty.

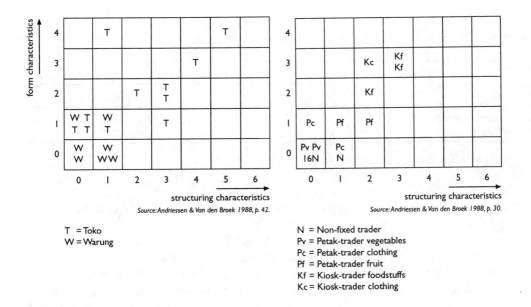

Figure 8.3a: Matrix position of toko and warung enterprises from the sample

Figure 8.3b: Matrix position of trade enterpises from the sample

Although toko's on the average are more often of a corporate type than warungs (informal retail-establishments) or pasar-traders, there are some notable exceptions to the rule. Especially the smaller toko's frequently are making use of informal sources of credit and (unpaid) family labour, thus displaying more petty than corporate characteristics. On the other hand, pasar traders selling garments, cloth and utensils on fixed plots (*petak*) or in a kiosk, frequently have as many corporate characteristics as the smaller toko's. A similar differentiation has been found among wholesalers (*tengkulak*), transporters, and money-lenders, thus invalidating the rigid bazaar-type versus firm-type dichotomy hypothesized by Geertz (1965), McGee (1971) and Missen (1972) in their earlier work.

On a higher level of analysis however, it is still possible to generalize about the dominant corporate or petty character of these various activities in order to study their complex interrelations and find an answer to questions of conservation and dissolution among the petty-producers.

On the average the petty-traders and producers were indeed found to be much more dependent on inputs from the corporate sector than the other way round. Many petty-producers however, do not buy from corporate enterprises at all, while others may be completely dependent upon them (e.g. cigarette or ice-hawkers). On the other hand, some corporate firms seem to lean heavily on inputs from petty- and peasant-producers such as cash-crops (tobacco, rice), building materials and processed food (*tahu, krupuk*), while most of them only sell their produce or services to petty-producers.

Table 8.2: Estimated percentage distribution of origin of inputs according to type of enterprise in Wonosobo

Supplier	Type of enterprise interviewed	
	Petty	Corporate
Petty enterprise	50	20
Corporate	50	20

Sources: De Jong and Ligthart, 1986, p. 141

The fact that petty-producers constitute quite a large part of the corporate clientele explains also why corporate firms will often refrain from taking over petty-enterprises. A clear example for this conserving type of interdependency is found in retailing between wholesale-distributors (*grosir*), shop-keepers (*toko*), *warung*keepers and *pasar*traders.

For shopkeepers an important reason not to compete on retail-level with petty-traders is that selling to consumers is less profitable, as in that case only small amounts and hence smaller turnovers can be handled. Usually commodities are supplied by the higher level (e.g. grosir or toko) to the lower level, frequently by extending credit, thereby reinforcing the dependency of the latter and increasing the prices at the end of the distributive chain. Another reason for the obvious lack of competition between corporate en petty-traders, is their specialization on different consumer-goods and customers. The shops usually serving the urban and rural middle classes with luxury items, whereas the warung and pasar-traders primarily are serving the poorer mass of the population with cheaper goods.

Similar symbiotic relations between corporate and petty-enterprises have been found also in construction-industry and transport-services through practices like subcontracting and leasing, thus invalidating the earlier hypothesis by Geertz 1965, McGee (1971) and others, that the expansion of the firm-type (*toko*) economy would inevitably drive out the bazaar-type (*pasar*) activities. It should be stressed however, that even under circumstances of symbiotic relationships, the interdependency between the sectors rarely is an equal one. The continuous transfer of surplus-value from the petty to the corporate sector is closely reflected in the different rates of net-profit in both sectors, being on the average Rp. 34,000 per month in the petty-sector against Rp. 439,000 in the corporate-sector (De Jong en

Ligthart 1986, p. 142). Consequently, the petty-producers are only able to survive owing to longer working days at lower returns per hour, i.e. through increased self-exploitation of (family) labour.

Next to these partly symbiotic and partly exploitative relationships between corporate and petty-enterprises there are also clear examples of competition leading to the dissolution of the latter. The most striking case being the penetration of corporate factories and trading firms with ready-made garments, shoes and utensils on the local markets for petty-tailors, shoemakers and other industrial homeworkers. The gradual takeover of the cheaper markets for these commodities is partly due to shifts in consumer preferences, but also to the cheapness of mass-manufactured products. Nevertheless, many of the petty-producers affected show a remarkable resilience towards corporate-sector competition. Partly by shifting their production to e.g. special dressings or shoerepair, but also by innovating their production methods or by increasing self-exploitation of labour.

The fiercest competition however, is not found between corporate and petty-enterprises but rather between enterprises of the same type and level, such as between toko's and bigger pasar-traders, whole-salers within or outside the region, and last but not least among petty-producers themselves. Especially the latter type of producers are suffering from supply-push induced growth tendencies within a limited and relatively poor market. A clear illustration of this is offered by the involutionary growth of petty rice-traders who in the various towns have been increasing faster than the total population or the trade-volume itself (de Jong & Ligthart 1986, p. 104).

Moreover, competition between petty enterprises in the commercial sector, frequently is countered by extending favourable credit-facilities to the customers, thus enhancing their cash-problems in times of shrinking markets.

It is remarkable however, that competition among petty-producers rarely is a cause for their dissolution. The main reason for this probably is that these activities frequently are of an additional kind, only supplementing the household incomes so that decreasing returns have less impact, considering also the low level of investment and the fact that losses often can be easily compensated for by making longer working days.

On the other hand competition between corporate enterprises may have much more serious effects if markets are shrinking and returns to investment are declining. This applies particularly to the corporate manufacturing and construction enterprises which in our research-towns are mainly supplying local and regional markets and are heavily relying on obsolete technologies and capital from formal sources of credit (contrary to e.g. the *toko*). More specifically it were the food and beverage industries, the cigarette factory and the local contractors in construction that were suffering most from the external competition of multinationals and bigger domestic enterprises. Several among these enterprises, like the coconut oil factory in Banjarnegara, and the noodle (*mie*) factory in Purwojero-Klampok were kept open only because of the amount of capital invested and the social consequences to the workers employed.

Another remarkable fact about the position of corporate manufacturing enterprises at the local (town) level is, that they hardly maintain functional relations with local petty-producers. Although most enterprises do use local raw materials, these are rarely procured through local petty-traders, nor is there much sub-contracting to local petty-producers. These manufacturing enterprises therefore, are operating in a relatively isolated position, contributing to the regional income, but with very few forward and backward linkages to other enterprises.

An aspect hitherto insufficiently recognized in the conservation-dissolution discussion is the generation of new types of petty-production as a result of corporate-sector development. In our research towns many examples were found of corporate firms subcontracting to own account-workers, thus leaving an important part of the risk-taking to these petty-producers. Many independent foremen, electricians and carpenters thus are engaged as subcontractors by bigger construction firms. Other examples being small car-repair shops (*bengkel*) depending on corporate trucking or minibus (*colt*) firms, drivers and their assistants (*manol*) depending on colt-renting firms, and carrymen (*pesuruh*) working for shops. The expanding activities of these corporate firms therefore, have a direct beneficial effect on the opportunities for this type of petty-producers. Corporate firms may also generate petty activities in an indirect way i.e. by selling consumer durables like motorbikes, t.v.-sets and watches, which require skilled repair facilities. Besides, even traditional food-vendors have found new opportunities in serving the increasing number of visitors and workers at bus-terminals, government offices and other institutions (schools). It has been estimated that in Wonosobo some 38% of the petty-enterprises have come into existence during the last decade when corporate enterprises have been expanding too. Thus demonstrating that the innovation and conservation effects of the functional relationships between the two sectors are at least off-setting the dissolution effects, if not more than that.

A closer look at the connection between the petty or corporate character of enterprises and their successfulness in terms of continuity and profit rates however, has shown that corporate enterprises indeed have a considerably higher chance of being successful.

Table 8.3: Successful enterprises in Purworejo-Klampok per sector and type of activity

Activity	Petty (n=93)		Corporate (n=25)	
	number	%	number	%
Trade	10	22	4	57
Industry	4	11	3	38
Non-trade services	0	0	5	100
Total	14	15	12	48

Sources: Andriessen and Van den Broek, 1988, p. 106

The main reason for this being not only that corporate enterprises have better access to capital, technology and markets, but also that petty-enterprises have only very small chances of developing into corporate-enterprises because of their fierce mutual competition under marginal conditions. It has been observed more than once in Indonesia that new and promising enterprises with low initial barriers are very soon inhibited in their growth by increasing competition from imitators seeking a living in the same branch of activity (cf. Rietveld 1986).

Intersectoral Relations at the Regional Level
The basic model of the urban production structure set out above, may be extended into a more complex model including the rural hinterland of the towns (cf. Titus, de Jong en van Steenbergen 1986). As the process of capitalist penetration and transformation is also affecting the economic structure of these hinterlands, both rural farm- and non farm-activities may be transformed into more commercialized types of activity, producing for a market instead of serving only local subsistence needs and using capital inputs and wage labour instead of family labour. The rural economy therefore, may become structured along similar lines as the urban economy, i.e. on the distinction between a predominantly non-capitalistic peasant sector next to a more capitalistic commercial (farm) sector. Both sectors may include farm and non-farm activities, whereas the positioning of these activities in the matrix runs parallel to that of the urban sub-system by scoring their form and structuring characteristics.

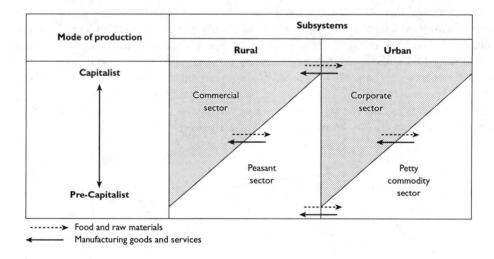

Figure 8.4: Model of the regional production structure and its intersectoral relations

The production relations between the various sectors in the urban and rural sub-systems are maintained by a continuous exchange of commodities, services (inclusive capital) and labour. In determining the type, direction and effects of these flows, the model leans heavily on earlier hypotheses formulated by McGee (1971) and Missen (1972) on the relations between the urban firm-type and bazaar-type economies on the one hand, and the rural peasant economy on the other. Thus the petty-commodity sector is assumed to play a vital role in the supply of the urban centres with food and raw materials from the peasant sector, whereas commodities and services produced by the corporate sector frequently find their way to the rural economy through petty-distributors as well. The main reason for this being that they can work at lower overhead-costs, needing less capital and usually working more hours per day at lower returns per hour.

These intricate relations imply that constraints imposed on one of the two sectors by another dominant sector, will have repercussions on both of them. Thus, if petty-producers experience heavy competition from corporate enterprises or even are dissolved by them, this will automatically reduce market outlets and cheap supplies for the peasant sector. Likewise, through agricultural commercialization and growing surpluses, peasant-farmers will increasingly be approached by corporate traders, taking advantage of the increasing economies of scale and supplanting the petty-traders. Moreover, increasing incomes among the emerging rural middle class may shift consumer preferences towards luxury items, like motorbikes, tv-sets and furniture, which are not suited to petty-trade because of their size and capital risks involved. Consequently, the petty-commodity network of rural produce marketing and distributive trade is becoming erodes, both by external competition and internal restructuring.

Illustrations of this were found in traditional tobacco (*garangan*) trade, as well as in the trade of cloves in Banjarnegara and the trade of palm-sugar in Purworejo-Klampok. In the uplands of Wonosobo tobacco is mostly grown by peasants (*tani*). After cutting and drying it is usually sold without any further processing to rural based petty-traders (*bakul*), who resell the tobacco to corporate wholesalers (*tengkulak*) in town (see figure 8.5a). More common however, tobacco-farmers have become commercialized farmers who are selling directly to the tengkulak, thereby short-circuiting the bakul-traders. The tengkulaks in their turn sell to factories outside the region, or to retailers in town (figure 8.5b).

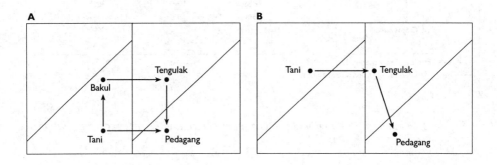

Figure 8.5: Two possible ways of tobacco marketing in the hinterland of Wonosobo

A discussion of the changing collecting-trade system at the regional level is not complete without analyzing the role of the KUD village cooperative in rice-trade. These village cooperatives function as part of the government's rice-marketing board and channel some 25-30% of the total rice production in the kabupaten Banjarnegara (de Jong & Steenbergen, 1987, p. 131). As each *kecamatan* (district) centre now has its own KUD rice-mill and storagehouse, this has considerably reduced the role of the corporate rice-traders in Banjarnegara-town, formerly monopolizing the milling and wholesaling of rice in the region.

Following this last example Titus, de Jong and van Steenbergen (1986) emphasize that rural-urban relations are not exclusively determined at the regional level. The impact of external relations on higher scale-levels should be considered as well, that is especially the role of corporate functions of higher order centres outside the region and the role of government policies at the national level. This again may be illustrated by studying town-hinterland relations and the marketing system at higher scale-levels in particular.

Town-Hinterland and Interurban Relations at the National Level
The structure of the peasant-marketing system in Java already has been described in detail by Dewey (1963) and Chandler (1984). Roughly a hierarchy may be discerned consisting of village (*desa*) markets, town-markets and city-markets connecting the local village-economy with the national economy. These markets have important functions in the bulking of commodities for regional and interregional trade, as well as in the breaking down of certain goods for local consumption an production. Most of the trade is within the local circuit, i.e. between and within villages, and is handled by petty-traders, whereas urban-based wholesalers are handling the upward and downward flows of goods at the higher (regional or national) levels. The upward flows consisting mainly of agricultural produce, whereas the downward flows consist of manufactured goods and inputs. Graphically the various circuits and levels may be represented as follows:

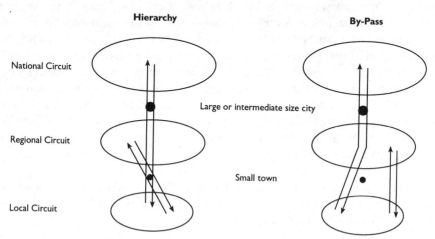

Figure 8.6: Commodity flows according to the hierarchical pattern and the by-pass pattern

At each level of these market-circuits however, the upward or downward flow may be interrupted by so-called 'by-passing' effects, thus inflicting a loss of functions and employment opportunities to the market-centres affected. Both Anderson (1980) and Mai (1984) in more recent articles, have pointed at the declining role of regional and local trade-centres in West Java and the Minahasa (N. Sulawesi) inspire of rising living standards and increasing populations. The main reasons for this being the increasing pressure on these markets from the national circuit and higher-order centres with the improvement of roads and transport facilities, the increase of agricultural surpluses and the shift in demand for manufactured goods. Wholesalers and their agents nowadays buy directly from peasants, thereby limiting the bulking phase to one step, whereas government-sponsored village-cooperatives (KUD) have the same effect on the trade in cloves and rice. Likewise, the Serayu Valley towns now are frequently being by-passed by corporate wholesalers from cities like Semarang, Magelang or Purwokerto, coming directly with trucks to buy vegetables, tobacco, cloves and palm-sugar from the peasant-producers. Consequently, the rural and small-town markets have become more of a buyers than of a sellers market, and the originally important collecting-trade functions of rural service centres like Banjarnegara is progressively erodes. On the other hand however, the distribution of manufactured goods neither follows the hierarchical pattern as described by Dewey. Increasingly these manufactured goods (consumer durables, textiles, utensils, soap) are being distributed through a system of toko's and traders dependent on them. The toko's have a tight hold on the market because they offer a wider range of goods against lower prices than the rural market-traders do. Moreover, they can render better services to the client, so that villages are inclined to buy in the toko of the district-town rather than on the local village market. Sometimes however, even here some by-passing of the service-centres occurs when sales-agents (*grosir*) from corporate enterprises in higher-order centres directly try to penetrate the rural markets by supplying local warung- and toko-traders with goods which they can sell on commission.

Another type of interference from externally based corporate enterprises in the regional system of town-hinterland relations, is by taking over specific activities in the regional service centres. In our research towns this was already obvious in certain branches of toko-trade (mainly in consumer durables and hardware) and in construction business. Usually this is done by turning the local corporate enterprise into a branch or subcontractor of the main company. Here it becomes obvious once more that these companies are not interested in taking over the petty-producers in these branches, partly because of their low turn-over rates, but also because many of the petty-producers are important customers or sub-contractors. Although this type of interference does not directly entail the by-passing of the regional service centres, it nevertheless may weaken their position by increasing their dependency on higher-order centres and by syphoning-off part of the surplus generated in these centres.

The close relationship between the small-town's production structure and their dependency on higher-order centres is reflected in the origin of the main types of commodities sold on the markets of Wonosobo and Banjarnegara.

Table 8.4: Percentage of traders at main markets in Wonosobo and Banjarnegara according to origin of main commodity sold

Origin of commodity	Agriculture		Home Industry		Manufacturing		Total	
	Wsb	Bnj	Wsb	Bnj	Wsb	Bnj	Wsb	Bnj
Inside kabupaten	29	37	21	24	1	-	51	61
Serayu region	5	3	3	4	1	-	8	9
Outside Serayu region Region	10	3	3	2	29	24	41	30

Sources: De Jong and Van Steenbergen, 1987, p. 76; De Jong and Ligthart 1986, p. 55

From this table it becomes clear that especially manufactured goods have to be brought in from other regions or even from abroad, because of the weakly developed manufacturing sector in these centres. Another interesting observation is that a more dynamic trade-centre like Wonosobo with its larger and more commercialized hinterland also offers a larger share and range of commodities from external origin than Banjarnegara.

If the interactions between the small-town economy and that of larger urban centres are represented in a schematical way, we may draw the following diagram (figure 8.7) showing that interaction remains confined to the corporate sector of the larger cities. The main reason for this being that competition from petty-producers in big cities is not likely, because they have higher operating costs (prices and wages being higher in metropolitan areas) and, in addition, transport costs for a fractionalized output will be too high (cf Santos 1979).

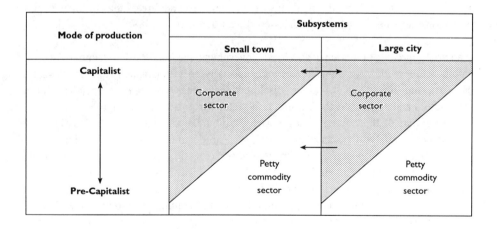

Figure 8.7: A sectoral model of inter-urban relations at the national level

The Role of Government Institutions and Policies
In the preceding paragraphs only occasional attention has been paid to the important role of government institutions and policies in shaping the small town's production structure and economic functions. Both in terms of employment and operational scale government institutions may be treated as part of the corporate sector, but this does little justice to the fact that they do not function as capitalist enterprises, nor as autonomous production units competing for a market. Government policies usually are formulated at the national level and implemented through local institutions at lower levels of administration, so that interactions between government institutions and the local or regional enterprises usually are of a unidirectional kind. Government policies and institutions therefore, should be treated as a separate element in our structural model of urban production relations.

Since the establishment of the New Order regime in 1967 the Indonesian government has attracted massive investments, developed new resources and expanded government control and institutions. Especially after the oil-boom in 1973 expanding government budgets have been used for increasing food production through Green Revolution programmes, expanding public services (administration, education and health), infrastructural facilities (roads, markets, irrigation works) and credit schemes for small-scale producers, but also for raising the salaries of civil servants. Additionally, the government became committed to the idea of regional planning in order to counteract the increasing social and regional disparities ensuing from its own capital and technology- biased development policy (cf. McAndrews 1982). Provincial planning boards (Bappeda I) were set up in each of the 26 provinces in 1974, followed by district-planning boards (Bappeda II) in every kabupaten-capital in 1980. Next to this a growth-centre strategy was adopted to encourage regional economic development. Most of these centres are intermediate or big cities (Cilacap, Semarang, Surabaya, Jakarta) and the results were disappointing, because trickle-down effects to small towns and rural areas have hardly occurred (cf. Wong 1984). As to the effects of these policies on the structure and functioning of small towns in our research-area, the following general observations have been made by Andriessen & van den Broek (1988):

a The strengthening of regional and local government has increased the importance of the public sector within the small towns and especially so in kabupaten capitals. Research in Banjarnegara and Wonosobo has disclosed that local government spending attracted contractors and stimulated the construction sector for public works (figure 8.8).
A certain part of these funds is appropriated by bureaucrats in a way that reminds one of Schatzberg's analysis (Schatzberg, 1979). These civil servants sometimes become entrepreneurs themselves, setting up enterprises (e.g. in photo-copying and ice-making) with the use of bank loans that are accessible because of their secure incomes.
b The increase in government expenditures has not only stimulated corporate enterprise activities, but also employment opportunities in petty-commodity production and services. The increase in salaries of civil servants for example is largely spent on buying extra food on the markets and paying for informal public and personal services (transport, domestic servants).

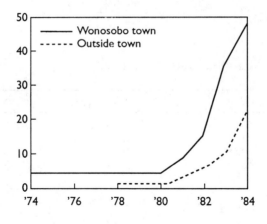

Source: De Jong & Ligthart 1986, p. 76.

Figure 8.8: Registered contractors in Kabupaten Wonosobo, 1974-1984

Still more important however, is the indirect effect of increasing government spending on petty-producers by increasing the total circulation of money in the urban and regional economies. It has been estimated that about one-third of each unit spent by the local government is directly absorbed by informal or petty-sector activities. Recently the government also has recognized the importance of the informal sector as a main absorber of surplus labour, since it has become evident that the capacity of the expanding formal sector in this respect is falling far behind expectations and that the agricultural sector already has reached its limits of absorption and even will decline in absolute numbers (Jones 1984). In order to stimulate non-agricultural activities the government has started special credit-schemes for small-scale entrepreneurs, such as the KCK and KIK small-investment credits (Hasibuan 1987).

c The growth of agricultural surplus has stimulated demand for local and non-local consumer goods in rural areas, especially the demand for consumer durables. Consequently, part of the rural surplus is syphoned off to the urban centres. Mobility of many rural dwellers has increased as well, because the number of minibuses for public transport has boomed during the 1970s. As a consequence the small town trade sector, especially the number of shops selling non-local consumer goods, has been expanding considerably.

d The improvement of rural roads that facilitated the boom in public transport, has also increased direct contacts between rural enterprises and non-local institutions, which entails that small towns are being by-passed. This process applies to collectioning as well as distributive commodity flows. In Banjarnegara and Wonosobo collectioning of rural produce still takes place, but by-passing seems to be increasing. This does not leave much hope for even smaller towns like Purworejo-Klampok, on which some authors nevertheless have put their hope for stemming the flow of migrants to the larger cities (cf. Hariri Hadi 1984).

Table 8.5: Origin of customers at selected shops in Wonosobo

Items sold	Town	Rest kabupaten	Outside kabupaten
Motors cars	36.0%	41.6%	22.0%
Motorbikes	41.3%	48.7%	10.0%
Furniture	63.0%	28.0%	9.0%
Smaller consumer Durables	38.5%	57.6%	4.2%

Source: De Jong and Ligthart, 1986, p. 121

The Urban Employment Structure and Labour Absorption

The main aim of this section is to analyze the small town's labour market in relation to its economic, social and administrative functioning. Central issues are the structure and functioning of the small town labour market, the opportunities it offers for the absorption of labour and income raising, and the dynamics it shows with respect to occupational mobility. By analyzing and comparing the results of our research into the employment structures of the three distinct towns Banjarnegara, Wonosobo and Purbalingga, we hope to reveal some of the main determinants and mechanisms in their functioning.

The small towns' employment structures and labour absorption conditions have been analyzed by making use of the urban labour market model as developed by Titus (1985) and presented in chapter 2. This model is operationalised and tested for its applicability in the next sub-sections.

The Small Towns Labour Force

Labour force[1] participation rates in the research towns generally are high (table 8.6). The high participation rates are explained by the generally low income levels, the absence of social security, and hence the need to obtain some kind of income, albeit usually a small one. This need is further aggravated by the young age composition of the sample populations in combination with prolonged school attendance and consequently the high dependency ratios.

The figures in table 8.6 show high participation rates, in particular for women[2]. This is partly explained by Javanese culture and partly by poverty. Within Javanese culture men and women are equally responsible for the economic support of their family (Koentjaraningrat 1967). Besides women take care of most of the material affairs concerning the household since they are considered more connected to the social and material environment and less able to deal with spiritual matters than men (Djajadiningrat-Nieuwenhuis 1987). The combined responsibility of women for both income earning activities and caring tasks, compels them to activities in or near the home with flexible working hours, usually own account work. Women thus are often employed as small traders at the market, as owners of small shops in their neighbourhood, as taylors and as producers and sellers of all kinds of foodstuffs. Adding to the appeal of own account activities for women is the fact that female wageworkers usually receive lower wages compared to their male colleagues (Lin Lean Lim 1993, p. 208). In public service jobs women are clearly underrepresented. In transport, a rapidly growing sector, women are conspicuously absent. Thus, the most secure and best paid jobs are nearly all occupied by men.

Table 8.6: Labour force participation rates

	Banjarnegara	Purbalingga	Wonosobo	Urban Central Java
Labour participation	70%	71%	83%	61%
Male participation	90%	87%	91%	81%
Female participation	53%	55%	70%	43%
Dependency ratio per worker	1.7	2	1.5	n.a.

Source: Household surveys Banjarnegara (1988), Purbalingga (1986), Wonosobo (1986) and National population census 1980 (Sensus penduduk 1980).

Female labour force participation rates in Wonosobo are higher in comparison with Purbalingga and Banjarnegara. The larger numbers of foodprocessing and trade activities in Wonosobo town, as compared to Banjarnegara and Purbalingga may explain for much of these higher rates.

Unemployment rates recorded in our sample surveys are 11% in Banjarnegara, 15% in Purbalingga and 8% in Wonosobo. The majority of the unemployed were young first time jobseekers with a relatively high level of education i.e. senior high school or vocational training. They were looking for a job in accordance to their education. The unemployed often belonged to relatively prosperous households which were able to support them until they had found a suitable job.

The unemployment rates of the research towns are above the average estimates for urban Java in 1986, reported by Jones & Manning (1992). Our sample results deviate from these rates because our surveys were carried out at the end of the school year. Hence, many recent schoolleavers looking for a job were reported unemployed. The lower average unemployment rates reported by Jones & Manning, however, may also refer to lower unemployment rates in the large urban centers of Java. In accordance to our findings, they observe also the highest unemployment rates among the better educated youth.

The sectoral composition of the labour force of the three towns is presented in table 8.7.

It reveals that Geertz's (1963) observation on the importance of public service employment and trade for the small town economy still holds true for Banjarnegara and Wonosobo. Purbalingga is deviating in this respect, as industry, next to commerce provides for most of the local employment.

Table 8.7: Employment structure of Banjarnegara, Purbalingga and Wonosobo

	Banjarnegara	Purbalingga	Wonosobo
Agriculture	9.6%	4.7%	2.7%
Industry / manufacturing and construction	10.6%	20.9%	19.6%
Commerce	37.8%	43.7%	42.2%
Government	21.9%	15.8%	20.6%
Other services	19.7%	15.0%	14.8%

Source: Household surveys, 1986 and 1988

Due to the delineation of the research town a rather large share of Banjarnegara's labour force is engaged in agricultural activities. The low percentage share of the labour force of Banjarnegara engaged in manufacturing/industry and construction is due to the absence of large enterprises and the dominance of petty type activities in food processing, tailoring and building material industry.

In summary, employment in the research towns shows the following characteristics; Trade is the main source of employment. Next to trade, the government is the largest employer in Banjarnegara and Wonosobo. In Purbalingga however, industry is the second largest source of employment, whereas in Wonosobo industry is nearly as important as government employment. Labour force participation rates are high, even among women. Child labour, however, is scarcely mentioned[3]. Nevertheless, children helping their parents after school hours is a common phenomenon. The educational level of the economically active population is not very high, but the completely untrained are only a small minority (6%).

Operationalisation of the Labour Market Model
As explained before, the urban labour market model used as an analytical tool in this study distinguishes three different employment sectors and three levels of integration in each sector [without adhering to the a priori assumption that employment in one sector yields higher or lower levels of integration that employment in one of the other sectors]. Based on the organisational characteristics of the activities the model defines a corporate sector, a family sector and an individual sector (cf. Friedmann & Sullivan 1974; Titus 1985). The corporate sector consists of larger corporate type enterprises, the public sector and large family firms.

The inclusion of public services in this corporate sector is justified by both their formal bureaucratic organisation, their employment conditions and their functional relationships with the private enterprise sector in Indonesia. Overall, the corporate sector offers the most secure and best paid jobs, although there are notable exceptions at the low level of integration. Entrance to this sector is mostly dependent on formal education, but also on capital ownership and personal connections/networks.

The family sector consists of small traders, domestic servants, service enterprises and small craft workshops or household industries. The main feature of this sector is the familial mode of production and employment of family labour. Kinship ties determine the labour relations and the entrance to this sector, next to a modest amount of capital for the entrepreneurs in this sector.

Finally, the in individual sector consists of small independent producers and own-account workers. As far as these independent producers employ any labour, it concerns only unpaid (family) labour. Entrance to this sector is relatively easy for labour without capital or special skills, but overall income security also is the lowest, so that Friedman and Sullivan (1974) have referred to it as a 'waiting room' for those who aspire employment in the corporate sector. The labour force from the household sample survey in our three research towns has been classified into the labour market model according to the following criteria: organisational form of the enterprise or type of activity performed, income level, security of income, certainty of employment, involvement in maintaining the existing institutional order[4] and occupational status. With the exception of 'the organisational form of the enterprise', these criteria determine the level of integration. This is in contrast to van der Post (1988) who operationalises integration in a strictly economic sense and measures integration by income level only.

Educational level or skill, it is argued, measures only potential employment and not actual employment. However 'integration' defined as 'an underlying process that provides access to participation in the urban structure' as Titus (1985, pp. 104) does, justifies the use of criteria like involvement in maintaining the institutional order, level of education and skill and security of income. These criteria provide an indication of the access one has, a.o. through relational networks, to different employment opportunities. Next to that, these extra criteria provide the opportunity to distinguish the really marginalised who lack access to opportunities to improve their situation.

The level of integration of the labour force in the three research towns is determined by following Titus' interpretation with an emphasis on income level and security of income. In determining the level of integration, income and security of employment were two basic criteria that were applied in each of the three sectors. These two criteria were supplemented with relevant criteria for each sector. In order to determine the level of integration in the corporate sector, education and involvement in maintaining the institutional order were added. The integration level in the family sector, is determined by adding criteria on ownership of the enterprise and/or capital goods, and on education and skills. Finally, the level of integration in the individual sector is determined by adding criteria on the ownership of the means of production, skills, education and investments to the basic criteria on income[5] and security of employment. In determining the sector and the level of integration of the labour force, the unemployed are also taken into account. Their educational level and/or skills, and therefore their potential access to either one of the sectors, determines to which sector they are assigned. They are included in the lowest integration levels.

Table 8.8 shows the results of the classification of the labour force of Banjarnegara, Purbalingga and Wonosobo. The labour force population is quite evenly spread over the three sectors of the model, while distribution over the integration levels is far from equal, with the largest share of the labour force at the lowest integration levels.

The corporate sector consists mainly of government employment. In this sector government and non-government managers are classified at the high integration level. Low rank clerical workers and factory labourers etc. are classified at the low integration level of this sector. Schoolteachers, and middle level public servants are classified into the medium integration level of the corporate sector. Employment in the family sector consists mainly of trade, ranging from petty type trade enterprises to corporate type shops. At the low and medium level of this sector owners of small scale manufacturing enterprises are found. They produce for example snacks, krupuk (shrimp crackers) or garments.

Table 8.8: Percentage distribution of the labour force of the three towns according to sector and integration level in percentages

Sector/ Integration level	Banjarnegara (n=415)			Purbalingga (n=451)			Wonosobo (n=715)		
	Corporate	Family	Individual	Corporate	Family	Individual	Corporate	Family	Individual
High	6	3	4	10	4	16	3	6	4
Middle	17	7	7	9	13	11	15	8	9
Low	14	21	23	9	18	9	16	22	17

Source: Household surveys

Common occupations at the low and medium levels of integration of the individual sector are small scale ambulant traders in foodstuffs, rishawdrivers, small scale markettraders, tailors and owners of small restaurants or shops. Market trade in clothes and household utensils, obviously occupations which require a reasonable investment and good supply and outlet relations, are found at the high integration level of the individual sector.

The larger share of the labourforce integrated into the corporate sector at the high and medium level in Banjarnagara compared to Wonosobo is largely due to activities related to the construction of the dam and the relatively greater overall importance of government employment in Banjarnegara. The relatively high percentage of the labourforce in Wonosobo town integrated at the highest level of the family sector is explained by the large number of toko's in Wonosobo town. The differences between the results for Purbalingga on the one hand, and Banjarnegara and Wonosobo on the other, are due to a different delineation of the income criterion and the presence of more corporate type factories and their staff. Furthermore, in determining the integration level of the labourforce of Purbalingga somewhat more weight was attached to the criterion of 'involvement in maintaining the institutional order'. The relatively large share of the sample population of Purbalingga classified into the high level of the individual sector is due to a deviating emphasis on ownership of means of production and/or profits from earlier investments. Table 8.9 shows the average income level per cell and reveals the deviating interpretation of the income criterion in the Purbalingga case study. Consequently, we suggest to compare only the combined figures of the two highest integration levels of this town.

Table 8.9: Mean individual incomes (x Rp 1,000) per month per cell of the model in Banjarnegara, Purbalingga and Wonosobo

Sector/Integration level	Banjarnegara			Purbalingga			Wonosobo		
	Corporate	Family	Individual	Corporate	Family	Individual	Corporate	Family	Individual
High	326	441	110	151	360	53	207	233	176
Medium	98	80	50	92	48	32	92	73	54
Low	39	11	23	42	11	9	33	23	24

Sources: Household surveys Purbalingga and Wonosobo 1986, Banjarnegara 1988

The high average incomes at the high integration level of the family sector are due to toko-owners and successful entrepreneurs who are classified into this segment. Their incomes however, are somewhat flattered because unpaid family work of their wives and children often contributes considerably to their income.

Differential Accessibility
Access to employment in the small town is regulated through three sets of variables. The first set refers to the demand for labour which is determined by the production structure and the government's policy with regard to investments in infrastructure and the appointment of government employees.

Demand for labour in the corporate sector consists of demand from the public sector and the few corporate type enterprises. Corporate type enterprises are only few, with almost no local backward relations and hence generating only limited employment outside the enterprise. Demand for labour or opportunities for income generating activities in the family or individual sectors therefore, depend to a great extent on the purchasing power of government employees, the amount of money circulating in the small town's economy through government expenditures and the purchasing power in the agricultural hinterland. Next to economic limitations, government regulations restrict demand for labour as well. For example *dokar* (horsecart) drivers in Wonosobo need a license from the local police office, a measure which enables the local government to control the number of dokars in town (v.d. Waal & v.d. Wouden 1988). Similar regulations control the number of minibuses. Furthermore, various informal exclusionary mechanisms[6] restrict access to income generating activities in the family or individual sector.

The second set of variables concerns the qualifications of the job-seeker. These are gender, education and skills. Gender affects the employment opportunities open to the individual. Certain types of jobs are still considered more appropriate to males like minibus driver or carpenter etc., while other jobs are considered more appropriate to women like teaching, producing and selling of snacks or domestic servant. Moreover, the urge for women to combine household tasks with income earning activities severely restricts their opportunities for better qualified, full-time jobs.

Education is essential to gain access to employment in the corporate sector, especially in the government bureaucracy. When comparing the educational level of the economically active respondents in the different segments of the model, education seems of minor importance in gaining access to employment in both the family and individual type sectors.

Access to employment, however, is not only determined by the individual's characteristics but also by the social economic status of the household to which he or she belongs. For often the household is a source of capital, capital goods, educational opportunities, relations and skills. Especially among occupations like shopkeeper, markettrader, farmer and artisan, it is general practice that children succeed their parents (v.d. Waal & v.d. Wouden 1988). The household thus determines to a great extent the opportunities open to the individual members in the jobseeking process.

Table 8.10: Prevailing educational attainment at the various segments of the model

	Corporative	Family	Individual
High	$\geq$ Senior Highschool	$\geq$ Junior Highschool	$\leq$ Primary school
Medium	$\geq$ Senior Highschool	Primary school	$\leq$ Primary school
Low	$\leq$ Junior Highschool	$\leq$ Primary school	$\leq$ Primary school

Source: Household surveys

The third set of variables influencing access to employment opportunities are the allocation mechanisms. These mechanisms can be divided into formal regulations and informal ones. Formal allocation mechanisms are the employment office and newspaper advertisements. According to the results of the sample surveys these mechanisms are only important with regard to corporate sector employment. Most respondents in the sample surveys, however, acquired their job through informal employment allocation mechanisms. The main informal allocation mechanisms are, kinship and acquaintance relations, middle men, bribery and the creation of self employment. Kinship and acquaintance relations are important in getting access to employment in each of the three sectors. Nevertheless, kinship and acquaintance relations are crucial in gaining access to the family sector. With respect to the corporate sector, or in their terms the modern sector, Jones and Manning (1992 pp. 306) point at the continued importance of family and other contacts (*koneksi*) in successful job-search activities in the modern sector.

In the individual sector kinship and acquaintance relations are of great influence to the success of the individual entrepreneur. For these relations often determine whether one can establish profitable supply arrangements, find a source of initial capital, and a suitable location where one can practice his (her) business and have a suitable marketing outlet[7]. Bribery to obtain a certain job, usually in public service, is not uncommon[8]. The amount of slush money varies per job, depending on status, possibilities for extra income on the job etc. Furthermore, the sum of the bribery depends on the existing relation between the applicant and the one who is receiving the bribe. But even if one is prepared to pay a reasonable amount of slush money it is still important to have some connections, otherwise there is a considerable risk the applicant is wasting his/her money.

The last job allocating mechanism worth to be mentioned, is the creation of self employment. This mechanism is predominantly at work in the individual sector, as it is the only sector characterized by a relative ease of entry through low capital requirements.

When reviewing the foregoing description of the accessibility of the labourmarket, it becomes clear that in a situation of an overabundant supply of labour, education, skills and capital are important, but personal relations are absolutely necessary. Those without any valuable relations have a fair chance to become only marginally integrated. Connections in other urban areas, however, might present an alternative and result in migration.

Marginalisation and Poverty
The very existence of both formal and informal entrance restrictions and barriers to integration in a labour surplus economy, does not only secure the incomes of the "haves" but also cause the marginalisation of the labourforce that is denied access to the more productive types of employment. Following the theoretical assumptions and due to the applications of the income criterium, the marginalised and poor share of the labourforce is mainly found in the F-3, I-3 and I-2 cells. The classification of the economically active part of the sample population (table 8.8) shows that quite a large share[9] of the economically active population of the three towns is indeed integrated into the marginal segments. Fortunately, the share of the heads of households (who are the main responsible for income and status of the household), integrated into the marginal segments of the model is far smaller. Nevertheless, nearly one third of the households depends for its main source of income on low paid and insecure income earning activities, which hardly provide the household with sufficient income and the necessary assets to either maintain or improve its income situation.

As we have noticed before, the classification of the labourforce according to sector and integration level is based on a set of criteria. But how does this classification compare if we only take into account income? In order to establish the extent of marginalisation and poverty according to income level, the rice purchasing equivalent is a useful instrument. This measure provides the opportunity to take account of differences in the costs of living, due to a.o. transport costs and inflation, in a relatively easy way. It defines the minimum adequate income at a certain place and time through the rice price. An adequate income, according to this standard is defined as an income that is sufficient to buy the equivalent of at least 20 kg. of rice per person per month in rural areas and the equivalent of 30 kg. of rice in urban areas (Sayogyo as quoted in Moir 1977). The larger amount of rice equivalents for urban areas expresses the urbanites dependence upon the market for their daily needs. From the rice purchasing equivalent we can calculate the minimum adequate income level for one person, for one person plus the average number dependents and for an average sized household[10]. The minimum income levels according to the Sayogyo norm are presented in annex 8.2.

When applying the Sayogyo income norm it turns out that in Banjarnegara town some 15% of the labourforce are unpaid family workers and another 13% earns less than the minimum required for a single person. The income situation in 1986 in Purbalingga and Wonosobo town is slightly better. In Purbalingga and Wonosobo town, respectively 11% and 9% of the labourforce are unpaid family workers. Moreover, in Purbalingga town 10% of the economically actives earns less than the minimum required for one person, while this figure is only 7% in Wonosobo town. These percentage shares of the labourforce earning less than the Sajogyo norm to sustain a single person, however, still represent a too large share of the labourforce that has to survive on the bare minimum of existence.

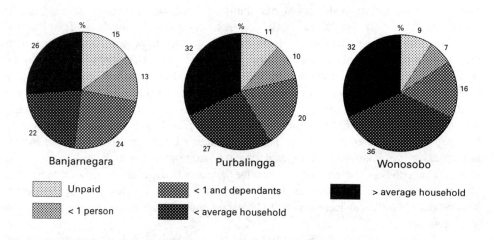

Figure 8.9: *The labour force according to earning capacity based on the Sayogyo norm*
Source: Household surveys

The shares of the economic active labourforce in Banjarnegara, Purbalingga and Wonosobo who do not earn enough to provide for themselves and the average number of dependents are respectively 24%, 20% and 17%. Hence the income situation in the research towns is rather poor.

Compared to the results of the 1976 labour utilization survey (Moir a.o. 1977), however, the situation has improved. The findings of this survey showed that 38.8% of the urban labourforce of Java earned less than the minimal necessary income for one person, whereas the income level of 78.8% of the urban labourforce was below the amount needed to provide for a single household. Still the income situation in the research towns compares very poor with the CBS (BPS) estimate of urban poverty in Central Java. The CBS applies an income criterion of Rp.17,381 per capita and estimates that 17.9% of the urban population in Central Java earns in 1987 an income below this poverty line (Firdausy 1994). In Wonosobo town 25% of the economic active population in 1986 earned less than the CBS poverty criterium, while in Banjarnegara town even 30% earned less.

Marginal incomes are mainly found among domestic servants, unskilled labourers, casual construction workers, agricultural workers, shop assistants, transport labourers, tricycle drivers and horsecart drivers. In Banjarnegara and Wonosobo the marginal incomes with less than the Sayogyo norm for one person, are concentrated in the low integration level of the family sector, and in Purbalingga town in the low integration level of the individual sector. Insufficient incomes for one person and the average number of dependents concentrate at the low integration level of the individual sector and in Wonosobo at the low integration level of the corporate sector as well. Insufficient incomes to sustain an average sized household prevail both at the medium level of the individual and family sectors and at the low integration level of the corporate sector while in Purbalingga town these low income levels prevail also at the high level of integration in the individual sector.

In a situation where on average two persons per household are economically active, real impoverishment occurs mainly among those who do not earn enough to provide a living for themselves and the average number of dependents.

The differences in income levels between Banjarnegara on the one hand and Purbalingga and Wonosobo on the other, could be due to the relatively large share of agricultural workers and farmers living in Banjarnegara town, since urban workers in the agricultural sector make up for 24% of the urban poor in Indonesia (Firdausy 1994). However, insufficient adjustment of income levels in the region to the rising price levels in the 1986-1988 period, caused a.o. by the $\pm$ 45% devaluation of the Rupiah in september 1986, combined with the influence of budget cuts in government expenditures may also explain for the unfavourable income situation in Banjarnegara.

Obviously the income situation of the small towns is rather precarious. This forces households to commit more labour to income earning activities than might be desirable from social and educational points of view. Furthermore, it must add to the attractivity of migration to the larger cities. It certainly is not a sign of greater dynamism of the small town's economy vis à vis the rural economy, nor of the small town as an alternative destination for rural-urban migrants (cf. Hariri Hadi 1984).

Occupational Mobility
Of course the hitherto analyzed situation is not a static one. Constantly newcomers enter the labourmarket, elderly retire, entrepreneurs go bankrupt, other thrive, civil servants make promotions, get transferred, etcetera. Occupational mobility patterns have been examined in order to estimate, whether or not the small town labour market provides adequate opportunities to the economically active to improve their situation.

Occupational mobility is defined here, as any change of job, occupation, function or activity. Within the framework of our labourmarket model, these changes lead, if they are considerable enough, to improved or deteriorated integration. A sketchy outline of the occupational mobility of the small town labourforce is obtained by comparing the present occupations of the economically active population with their first jobs. In particular with regard to women this approach might underestimate their occupational mobility. For, rural Javanese women tend to change occupation in order to adapt themselves to the changing claims of their family for care (Peluso 1984). The caring tasks of women change over time along with the birth of children, the ageing of the children and the demands for care by the parents. Women usually change occupation when they get married or just before their first child is born. This change often concerns a shift from activities outside their house to an activity near home, for example a change from streetvendor to *warung* (small shop) owner. As the children grow older, the women start to spend more time in income earning activities and spend longer hours away from home. Ageing women tend to spend more time at home and leave more of the 'outdoor' activities to their children.

The comparison of the first and the present occupations of the sample populations yields an overall picture of occupational stability. Of the economically active sample population 51% in Banjarnegara never changed occupation, job or function. For Purbalingga and Wonosobo these figures are respectively 63% and 54%. The observed occupational stability is the result of such different factors as, population pressure which causes severe competition for income earning activities, the existing relations between employer and employee, the importance of supply and outlet relations in own-account work and the lack of investment capital. The relation between employee and employer is in general informal, instead of business-like. Towards his employees a good employer is assumed to act more or less like a father, a role which inhibits him (her) from firing employees all too easily (Hofstede 1985). An example of this kind of labour relations is presented by a mie (noodles) factory in Banjarnegara. The factory is operating at a loss but for the sake of employment is not closed down. The Chinese owner covers the losses from profits made in his other business enterprises (Verbeek & Schillemans 1989). Own-account workers, put a lot of energy and time in the development and extension of their supply and marketing network. These relations are often reinforced through the extension of credit. A change in trade or kind of goods produced therefore involves an enormous investment because it necessitates not only a change in knowledge about the commodities but also the establishment of an entire new network for supply and outlet.

Following the model, our analysis focuses on occupational mobility resulting in further integration or in marginalisation.[11] The majority of the labourforce has never changed jobs in such a way that it has resulted in a shift of position in our labour market model.

Table 8.11: Percentage shares of the economically active who never changed position within the model, per segment

	Corporate Sector			Family sector			Individual sector		
	Banjar-negara *)	Purba-lingga	Wono-sobo **)	Banjar-negara	Purba-lingga	Wono-sobo	Banjar-negara	Purba lingga	Wono-sobo
High	22%	48%	30%	33%	60%	74%	73%	53%	44%
Medium	79%	64%	79%	48%	57%	62%	63%	57%	66%
Low	61%	78%	79%	86%	89%	53%	79%	56%	75%

* Figures are based on 358 respondents of whom both sector and integration level of their present and first occupation are known.

** Figures are based on the first and present occupations of 275 economically active heads of households.

Source: Household surveys

Table 8.12: Labour mobility within the model compared

	Banjarnegara	Purbalingga	Wonosobo
Corporate sector	n=121	n=128	n=108
* Unchanged	64%	69%	69%
* Upward, from within the sector	15%	17%	17%
* Downward within the sector	3%	4%	-
* Inflow from the family and the individual sector.	18%	9%	15%
Family sector	n=118	n=159	n=89
* Unchanged	75%	73%	62%
* Upward, from within the sector	13%	11%	8%
* Downward within the sector	-	1%	1%
* Inflow from the corporate and the individual sector.	12%	15%	29%
Individual sector	n=119	n=156	n=78
* Unchanged	75%	56%	63%
* Upward, from within the sector	8%	7%	4%
* Downward within the sector	-	-	1%
* Inflow from the corporate and the family sector	15%	37%	32%
* Unknown	2%		

Source: Household surveys

The figures in table 8.11 clearly show an overall stability. Only at the high integration level of the corporate sector the majority started their careers in one of the other segments of the model. The majority of the economically actives, now integrated at the high integration level of the corporate sector entered the labourmarket either at the medium or at the low level of the corporate sector.

This is due to the fact that most of the employment in the corporate sector is within government service, which is organised in a strictly hierarchical way and is characterized by frequent promotions and changes of function. Moreover, entry to the government sector usually is blocked by various types of credentials, connections and fees, which make it a more or less closed shop.

Table 8.12 presents a comparison of processes of integration and marginalisation within and between distinct sectors of the labourmarkets of our three towns. In line with the figures presented in table 8.11, the majority of the sample populations did not experience further integration or marginalisation. Further integration in the corporate sector mainly takes place between the medium and high integration levels (Banjarnegara 72%, Purbalingga 55% and Wonosobo 61% of the integration within the sector). Marginalisation within the corporate sector had affected only a small number of respondents. It occurs mainly between the medium and low levels of integration. The influx of labour from the family and individual sectors predominantly originates from the low level of the family sector. They enter the corporate sector either at the low or at the medium integration level. In the flow of labour from the family sector to the corporate sector the young and well educated prevail. They often take up work in a family enterprise while searching for a job in the civil service. Integration in the family sector predominantly occurs in the low and medium integration levels. It often concerns formerly unpaid family labourers who took over the business from their parents. Quite remarkably further marginalisation in the family sector is hardly measured, thus illustrating the relative stability of the employment relations in this sector.

The influx of labour to the family sector in Purbalingga and Wonosobo mainly originates from the low integration level of the corporate sector (50% of the inflow in Purbalingga and 46% in Wonosobo). This flow concerns labour from private formal enterprises who prefer an insecure income in a family enterprise over a secure but low paid low status job at the low level of the corporate sector. Labour originating from the medium level of the corporate sector predominantly concerns former civil servants who after their retirement at the age of 55, start their own business.[12] In Banjarnegara town the influx of labour to the family sector comes predominantly from the low integration level of the individual sector, due to the relative shortage of private corporate type enterprises.

The three towns show a different pattern of integration in the individual sector. In Banjarnegara town labour mobility from the low level to the medium level of the individual sector prevails. In Purbalingga town labour movement prevails from the low and medium levels of the individual sector. Whereas in Wonosobo town integration processes within the individual sector are of minor importance. Further marginalisation within the individual sector is hardly measured in either of the three towns. However, influx from the family and corporate sectors occurs at all three integration levels of the individual sector. Labour originates from the low and medium levels of the family and corporate sector. In Banjarnegara town the inflow of labour concerns mainly a flow between the low level of the family sector and the low integration level of the individual level. This flow consists predominantly of former family workers who have acquired enough skills, experience, relations and capital to start for themselves. In Purbalingga town the inflow of labour to the high and medium levels of the individual sector mainly originates form the low levels of the family and corporate sector. Labour from these segments prefers self employment, profit opportunities and status over job security and low status low paid employment at the low integration levels of the family and corporate sectors. In Wonosobo town, labour originating from the low level of the corporate sector prevails in the flow to the individual sector. This flow consists of pensioners and former labourers from corporate type enterprises. To the pensioners own account work provides an opportunity to supplement their small pensions. Whereas to the former factory workers selfemployment offers, independence, the opportunity to earn the profits of their own work and sometimes higher status.

Occupational mobility between sectors mostly takes place at the lower levels of integration, and thereby confirms the assumption that it is a phenomenon at the base of the urban economy. More surprisingly, the mobility pattern found in the three small towns did not confirm the overall assumptions in the literature (a.o. Friedmann & Sullivan 1974). Based on the concept of a formal-informal sector dichotomy of the urban labourmarket and assuming a higher status ascribed to the formal sector, it is hypothesized that employment offered by the formal sector is far more attractive than employment in the informal sector. The informal sector is even considered a 'waiting room' for employment in the formal sector or 'a last resort' for those who can not obtain employment in the formal sector (Todaro 1985). Following this reasoning, there is a continuous flow of labour from the informal to the formal sector which is also more extensive than the return flow. When analyzing the mobility paths between the sectors in our samples, the flow of labour from the corporate sector to the family or the individual sector appeared to be of far greater importance than the flow from the individual and/or family sector to the corporate sector. The flow from the corporate sector to the family or the individual sector consists of retired civil servants and of employees of large private enterprises. The mobility pattern of the latter category, however, is only explained by a greater attractivity of own account work or family enterprises over secure but low paid wage-employment. The opportunity to earn a higher income as an own account worker than as an employee, probably adds to the attractivity of own account work, even if it is less secure and does not offer supplementing employment conditions, such as a free lunch, health insurance etc. It turns out that although labour productivity in larger scale enterprises is higher, this does not accrue to the labourers. In these circumstances it is often more attractive to start a private (single person) enterprise of which at least all the surplus produced accrues to the owner/worker. A similar trend in labour mobility has been observed by Teilheit-Waldorf & Waldorf (1983) in Bangkok, and by Blau (1986) for Malaysia. By comparing incomes of non-educated own account workers (or informal sector workers) and non-educated employees, both demonstrate that own-account workers earn on the average more than employees. Blau (1986) mentions for Malaysia a tendency for young people to start their career as an employee and turn to own account work when they grow older. This tendency has also been noticed in Purbalingga (Klapwijk 1988). In Banjarnegara and Wonosobo this pattern was less clear.

In brief, the overall picture of occupational mobility in the small town labourmarket is one of stability and stagnation. Integration usually takes place within one sector. Occupational mobility between sectors consists mainly of a mobility path from the corporate to the family and the individual enterprise sectors as a result of the relative attractiveness of own account work over wage employment in the corporate sector, and because retired civil servants supplement their meagre pensions with own account work or some sort of family enterprise.

If our analysis of the occupational mobility and employment opportunities in the research towns is representative for the overall pattern of urban occupational mobility, it may have far reaching implications for government policy. It implies a shift in policy emphasis away from large scale industrialisation to the support of small scale entrepreneurs. This shift involves among others the abandoning of restrictive measures towards small scale entrepreneurs, like transporters or traders, the improvement of the efficiency of government sponsored credit programmes and more generally, greater attention in government policy towards the needs and interests of family and individual enterprise workers.

Conclusions

In the preceding sections our structural models have demonstrated to be of considerable value for analyzing the urban production relations and functions at various scale-levels. First, it was shown that urban activities and enterprises indeed can be classified according to their dominant corporate or petty-commodity features, taking account of both their mode of production and form-characteristics. In this way it became possible to study the functional relations between the various types of enterprise or sectors, and analyze the ensuing processes of dissolution, conservation and innovation.

A striking result of this type of analysis is that the relations between corporate-enterprises and petty-producers at the local level often are more of a symbiotic (or at least neutral) nature than of a competitive nature. This applies especially to the trade and non-trade services, as well as to construction industry, where the higher level is making use of the lower level as a distributive outlet or as a subcontractor, sometimes binding the latter through credit-systems. In manufacturing however, the corporate sector does not entail much local spin-off through subcontracting or demand for inputs. The main sources of competition for local petty-producers, causing their dissolution are stemming from external corporate producers and traders, and not from local corporate enterprises. Thus, at the local level competition is most fierce between producers of the same type, serving the same market. Among petty-producers however, this rarely leading to their complete dissolution because of their flexibility through self-exploitation of labour and low overhead costs.

The net-effect of public and corporate sector-growth on petty-production in these small towns therefore, probably is a positive one. By increasing the local circulation of money and goods, both the public and corporate sectors create new opportunities for petty-producers, i.e. especially in transport, trade, repair and personal services. Tertiarization of the small-town economy in this respect does not necessarily imply a proof for its further involution, although the marginal position of most of these petty-services remains obvious.

A quite different picture emerges when we are observing the small-town's regional functions and relationships with its rural hinterland. The assumed role of these towns as important distributive and collectioning trade-centres was only partially confirmed by the facts, i.e. as far as the distributive functions are concerned. The main reasons for this being that processes of agricultural commercialization and surplus-production have induced farmers to rely on input and output relations with external wholesalers and suppliers, thus by-passing the service-centres and reducing the role of petty-traders as a liaison between the corporate and peasant sectors.

Moreover, policies and programmes aiming at agricultural and rural development have been largely determined by higher (national) levels of decision-taking, thus by-passing the local service-centres too. What is being left, is their role as distributive trade-centres, serving mainly the increasing demand of the emerging rural middle class (next to that of the urban middle class). This usually reinforces the position of the town's corporate sector enterprises, as this changing situation involves also a shift in demand from petty-products to manufactured consumer durables.

At the same time however, the small-town's corporate sector is experiencing increased competition from externally based corporate enterprises penetrating the local and regional markets with superior or cheaper products, and taking over the most successful branches. Thus, the potential role of the small-town as a regional growth centre (cf. Rondinelli & Evans 1983) is mainly reduced to that of a regional service-centre, following passively the developments and structural changes induced in its hinterland by external forces originating at the national level and in higher order centres.

The employment situation in the research towns is characterized by relatively high labour participation rates, rather low unemployment rates and a high incidence of poverty. Trade is the main source of employment in the research towns. Public services come second, except for Purbalingga town where industry/manufacturing comes second. The labourmarket in the research towns is saturated as supply of labour is abundant and demand is only limited. As a result, own account work is widespread and migration is a serious option. In this respect the employment situation is much more determined by supply instead of demand. When comparing the three towns it turns out that Wonosobo town is characterised by the highest labourforce participation rates, lowest unemployment rates, lowest dependency ratio and the lowest level of poverty. Although high labour participation rates, a low dependency ratio and low unemployment rates are interrelated, the labour absorbing capacity for trade in Wonosobo town, still compares better to Banjarnegara and Purbalingga, judging by poverty levels.

Differentiation in the employment situation in the three towns is not so much reflected in the occupational (sectoral) composition of the labourforce. It is more adequately reflected in the labourmarket model used here in analyzing the employment situation. The classification of the economically active according to the model for example showed much similarity between Banjarnegara and Wonosobo town. It holds for all three towns that a far too large share of the economically active is only integrated in the marginal cells, occupying the low income earning activities, with low job security and low educational and skill requirements. This picture becomes even worse if one takes into account that upward mobility from the low levels of integration hardly occurs, except for the corporate sector. More than half of the sample populations even has never changed sector and integration level. As compared with Banjarnegara and Wonosobo, the labourforce in Purbalingga is the best integrated.

This is partly explained by Purbalingga's more differentiated production structure and to the presence of corporate type factories in particular. The staff and management functions in these factories certainly contribute to the relatively high percentage share of the population integrated at the high level of the corporate sector. The relatively favourable integration record in the individual sector, however, results from the greater weight attached to criteria on the ownership of means of production and profits / proceeds from some sort of investment in combination with a less strict application of the income criterion. This resulted in an upgrading of the overall level of integration of the economically active in Purbalingga. The low average incomes in the respective cells of the model for Purbalingga confirm this. The relatively high percentage of the economic active in Wonosobo integrated at the high level of the family sector and the high average incomes found in this cell is directly related to the large and thriving toko sector in Wonosobo town.

These marginal employment conditions are reflected in the small towns' income situation. Nearly 25% of the sample populations earned less than the Sayogyo norm for one person and the average number of dependents. This adverse income situation explains for much of the high labourforce participation rates in the three towns.

Occupational mobility seems hardly to occur. As far as occupational mobility takes place, it is mainly confined to the corporate sector, which means in the small town context, to government services. Labour mobility between the sectors occurs mainly at the low level of integration. Within this mobility pattern the flow from the corporate sector to the family and the individual sectors prevails. This mobility pattern and the meagre average incomes earned at the low integration level of the corporate sector, call for an adjustment of the marginal segments in the model. As the flow from the corporate to the family and individual sectors consists of pensioners and employees of private enterprises, it would be justifiable to differentiate at least the C-3 cell into a corporate public and a corporate private cell. The existing mobility pattern and the low wages common in factory work justify the inclusion of this (corporate-private) cell into the marginal segments.

In the rather stagnant labour market situation it comes as no surprise that the research towns do not attract the mainstream of rural-urban migrants from the Serayu area. Labour migrants who did settle in the research towns are either civil servants from other parts of Java who have been transferred, or rural dwellers who work as domestic servants or small scale traders.

The stagnant labourmarket situation and the relatively high unemployment levels among young highschool graduates causes considerable outmigration from the research towns. Suitable employment for the highschool graduates is limited to demand from government institutions, since the few corporate type enterprises in the research towns add only very little to the demand for the higher educated. Hence, limited employment opportunities in the research towns are a major reason for many a migrant to leave.

On the whole, the resemblance in employment structure of the three research towns is striking. The dominant administrative function the towns perform causes a quite large number of government employees to reside in these towns. These civil servants with their fixed salaries will create a stable demand for commercial services. It is tenable to assume that this demand will be more or less equal in the three towns, because they perform the same function in the administrative hierarchy. Furthermore, a large share of the development funds from the national government, allocated to the district, is spend on infrastructural works thought officials and companies based in the district capitals. This adds to the amount of money that is circulating in these towns. In principle, this circulating capital might offer opportunities to small entrepreneurial activities as well. However, in the mid eighties the Indonesian government set out on a structural adjustment programme leading to severe budget cuts. Even at the end of the decade with the recovery of the Indonesian economy, it did not seem that growth in demand for locally produced products and services from large scale enterprises had substituted for these budget cuts. Moreover local backward linkages of enterprises are limited as we have argued in the preceding section. Hence demand for labour and opportunities for own account work will remain limited as well.

The yearly numbers of school leavers and the rural labour surplus, add up to the pressure upon the small town labour market and thus will continue to push people into petty type activities, i.e. in employment at the low integration level of the family and individual enterprise sectors. Part of this pressure upon the stagnant urban labour market however, is disbursed through a steady outflow of migrants from the young productive age groups as is demonstrated by the sudden dip in these age groups in all three towns (annex 8.3).

Notes

1. Labourforce in this context is defined as everyone of 10 years and over who no longer attends school with the exception of those who are not able to work because of sickness or old age (Bijlmer 1987).
2. The labourforce participation rates are well above the Indonesian average of 60% for males and 22% for females in 1986 (SAKERNAS quoted in Pernia & Wilson 1989). The SAKERNAS figures, however, do not include unpaid family workers and part-time labour working less than one hour.
3. According to estimates of the Indonesian Central Bureau of Statistics ± 10% of the children in the age of ten to fourteen years works. Around 80% of the working teenagers live and work in rural areas (Godenbauer & Trienes 1995).
4. The criterium of involvement in maintaining the institutional order is particularly applicable in the Indonesian context, because the government is represented at all levels of society. Good relations with the bureaucracy are a pre-requisite to improve or even consolidate one's economic and social position.
5. The income criterium was delineated as follows:

Sector/ Integration	Corporate sector	Family sector	Individual sector
High	> Rp.120,000	> Rp.90,000	> Rp.60,000
Medium	Rp.60,000-120,000	Rp.45,000-90,000	Rp.30,000 - 60,000
Low	< Rp.60,000	< Rp.45,000	< Rp.30,000

6. An example of an exclusionary mechanism is the way a manual and his friends preserve their income earning activity. A 'manual' arranges at a certain point a long the route of a minibus passengers and goods. The manual receives in return for this service a certain amount of money per passenger from the driver. To secure their income the manual and his friends make sure that no one else dares to start as a manual in near to them.
7. For example the wives of civil servants who engage in foodprocessing and find a steady outlet in catering for civil servants meetings.
8. The Banjarnegara research mentioned case of someone who offered a waterbuffeloe in exchange of a job at the watershed constructionside.
9. The percentages of the economically active sample population classified into the marginal segments of the model are for Banjarnegara, Purbalingga and Wonosobo, respectively 51%, 38% and 48%.
10. An averaged sized household, consists in Banjarnegara of 5.7 persons and in Purbalingga and Wonosobo of respectively 5.8 and 5.9 persons.
11. Integration refers to reaching a better level of integration, for example someone who starts as an unpaid family worker, than changes profession and becomes a government employee reaches a better integration level. Marginalisation refers to changes in the opposite direction, usually resulting in a further impoverishment.
12. In Indonesia civil servants retire at the age of 55. In order to secure their income after retirement they might set up a business while still in the public service. Others start a business after retiring to supplement their pension.

CHAPTER NINE
In the Shadow of Yogyakarta? Rural Service Centres and Rural Development in Bantul District

Henk Huisman and Wim Stoffers

In the early morning hours, the dense network of roads just south of the city of Yogyakarta offers an overwhelming scenery. On every strip of tar thousands of rural dwellers try to find their way to various destinations. Mainly heading to the (peri)-urban areas in the north, bicycles, carts, light-motorbikes, mini-vans, carriages, trucks, ox-wagons, pole-bearers, fight for room in a strikingly stoical way. Those who are foolish enough to attempt to toil up in the opposite direction are confronted with a road-wide smoky flood that behaves like an all-enveloping mass of things which is virtually impossible to negotiate. A truly claustrophobic sensation develops since everybody attempts to overtake everybody and the collective sham-deafness seems only matched by a collective virtuosity: accidents rarely occur and nobody shows any signs of stress.

Multiple exposure of rural geographers to such experiences, inevitably, leads to a number of questions. In the first place there are the obvious `what', `who' and `why' questions, or in full: for what purposes do all these rural dwellers travel? Which strata of the rural population are mostly represented on these roads? What do these particular groups actually want to sell, to buy, to obtain or to do in the big city and all these other destinations? Second there are questions which refer to the impact of the observed phenomenon, such as, for example, what do the massive daily flows to the city of Yogyakarta imply for the rural areas from which they originate? The present study will concentrate on these latter matters, i.e. on the influence of a nearby big-city on the rural services delivery system.

Until 1997, Indonesia has been undergoing a sustained high growth rate of the economy. Although external demand has been forming the basis of the process, internal demand has increasingly absorbed products and services. An important share of such demand has been of rural origin. Bantul district is an example of a densely populated rural area in the central part of Java that is experiencing a progressive transition from an agriculture-dominated economic structure to a more complex, diversified economy. Typically, outmigration remains limited and no signs of a rural exodus are present. On the one hand, this can be explained by the increase of non-farm income generating opportunities in the area itself. On the other, it is clear that the large number of cheap but excellent connections of the countryside with nearby Yogyakarta, and other centres of the Javanese urban system, allow rural household members to easily participate in commuting and other forms of circular migration, either or not in combination with agricultural or non-agricultural activities at home.

Obviously, this process of transformation has far-reaching consequences in terms of the rural settlements and their functions. Under these highly dynamic conditions, the functional structure of the various types of rural service centres for the population in their surrounding areas can be assumed to change rapidly as well. The large range of services present in the city can be expected to strongly influence the extent to which the rural service centres are used by the local population. Consequently, the presence or availability of services in these settlements will be affected. However, the impact will be differential: factors such as the centres' (relative) distance to Yogya and both the services' nature and level are likely to be of crucial importance in this respect.

The richness and the complexity of these issues are of such magnitude that a large quantity of empirical data is required. In a cooperation with the Gadjah Mada University in Yogyakarta and the regional planning board (BAPPEDA Bantul) we have attempted to build such a data base[1] (see also annexes 9.1 and 9.2 for a summary of methodological aspects and detailed quantitative information). Analysis of the available data allows for some insight into the actual role of small centres in the regional development process in this highly dynamic part of rural Java.[2] Main attention focuses on the assessment of the settlements' hierarchy and centrality on the basis of aspects related to the supply and use of services, respectively. The first is based on a combination of services present, while the second concentrates on the actual number of incoming interactions for a certain purpose. The structure of the text follows this line: subsequent to a brief outline of the main characteristics of the regional setting, the hierarchy of settlements will be presented on the basis of services provided. Next, attention focuses on the centrality of settlements on the basis of services used. In a case study on the use of community services attention will be paid to household related factors and their relevance for the use of these services. Finally, the factors considered relevant in the assessment of the role of small centres situated in the shadow of a big city are briefly discussed in the conclusion.

The Setting

The district of Bantul (507 square kilometres) is situated in densely populated Central Java, Indonesia, just south of the City of Yogyakarta. In 1990, at the start of our study the population amounted to 688,195 persons, implying a population density of 1357 persons per square kilometre (PEMDA Bantul 1991). In 1994, these figures had risen to 728,970 and 1438, respectively (PEMDA Bantul 1995). Since half of the area is used permanently as farmland, the agricultural density is about twice as high. The district enjoys a good access to the islands' major transportation routes, i.e. railways and roads. It can be subdivided in a number of distinct zones, viz. a rather flat lowland area in the central and southern part, a calciferous plateau in the western part, and a topographically rather rough upland area in the eastern part. The eastern upland area is clearly demarcated from the lowlands by a steep escarpment with slopes of 40 percent and over, which rises to an altitude of some 500 meters above sea level.

Some parts of this area are relatively inaccessible. The soils are largely lateritic and have a low degree of fertility and a limited moisture retaining capacity. The western plateau, which rises to an elevation of some 150 meters above sea level, also has soils with limited fertility and problematic moisture retaining capacity. In contrast to these parts, the central lowland zone - which comprises the main part of the district - is offering a highly valuable agricultural production potential. It has thick soils of volcanic origin that are highly fertile with favourable moisture retaining capacity levels. The topography of this zone allows for permanent irrigation of most of the farmland.

To do justice to this internal differentiation, a classification of the seventeen subdistricts in the district has been made on the basis of two criteria (see figure 9.1). First, the proportion of the work force active outside the agricultural sector is used to identify non-rural areas: subdistricts with more than 65 percent of the work force active outside agriculture have been earmarked as peri-urban areas. Second, the percentage of irrigated land per subdistrict is taken. Three separate categories can be discerned, viz. the dry land category with less than 10 percent of irrigated land (rural zone 1), the sawah dominated category with 36 percent of irrigated land and more (rural zone 3), and the category which falls in between these values (rural zone 2).

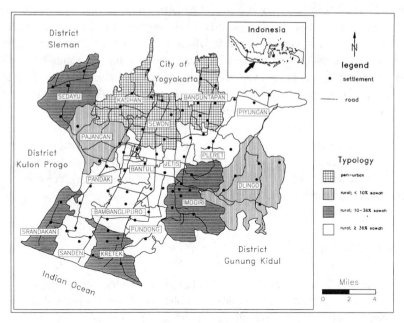

Figure 9.1: Typology of subdistricts in Bantul District
Source: Huisman and Stoffers, 1988

The agro-ecological differences are also clearly reflected in the distribution of the population over the district. In the eastern and western part, the population density is less than 600 per square kilometre - in the central part, figures are up to five times as high. Although agriculture still constitutes the most important single source of employment in most of the district large differences can be seen between the various zones. In the peri-urban parts near Yogyakarta, non-agricultural activities make up the main activity for more than 65 percent of the households (Huisman and Stoffers 1988 & 1990). In the other parts a close relationship exists between the agricultural system (dry or irrigated agriculture) and the involvement of the households in non-agricultural activities. In the wet-rice cultivation area land is a scarce resource and, consequently, the non-agricultural activities make up the main source of employment for almost half of the households; where dry land farming prevails, non-agricultural activities comprise a secondary source of employment only (ibid.). Non-agricultural activities encompass trade (11 percent of the total number of active persons), construction (7 percent), government services (9 percent) and manufacturing (8 percent) (PEMDA Bantul 1995).

The differences in population density are also reflected in the distribution of the settlements, according to size, over the area (figure 9.2). The larger settlements are concentrated in the northern part of the district, in fact they form the urban fringe of the City of Yogyakarta. Besides, the more densely populated central zone can clearly be distinguished from the less densely populated parts in the east and west. In the latter areas, the number of settlements, as well as their size, is far lower than in the first. In the administrative hierarchy, each subdistrict has one `capital', i.e. one settlement in which all government offices are located. The subdistrict capitals are not necessarily the largest settlements within the administrative unit.

The settlements' physical lay-out varies between the accidented and low lying parts. In the former zone, the built up area is dispersed; in the latter the built up area is much more condensed. This is related to the nature of the agricultural resource base and the relative value of land. In addition, the irrigation infrastructure greatly determines where new homesteads can be located. Settlements grow in a different way in the zones as distinguished: in the dry land area, the villages grow by spatial expansion; in the sawah zone growth by fission predominates. Obviously, this does not apply to the settlements bordering Yogyakarta city. In these parts both types of settlement growth can be observed.

Hierarchy and Centrality

In the assessment of both hierarchy and centrality of the settlements a number of aspects should be taken into account. First, the sheer number of service supply points in a centre cannot be used to determine its level in the hierarchy. The presence of a large number of primary schools, for instance, does not automatically imply that such settlement has an important educational function for the population in the surrounding area: it can simply be that many children in the relevant age bracket are present in the centre itself. In contrast to this, the number of inhabitants as such can be of minor importance to explain the presence of certain types of services. Often, the local cooperative (*KUD*), the branch of the Peoples Bank of Indonesia (*BRI*), and the rural health centre (*puskesmas*) are not located in the same settlement, although they are clearly meant to serve the same population. This shows that political considerations can have played crucial roles in the distribution of government services.

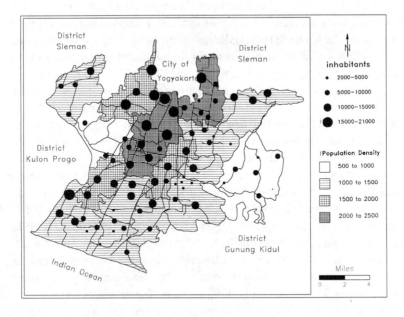

Figure 9.2: Settlement size and population density
Source: PEMDA Bantul 1991

Therefore, in the determination of the hierarchical level of a settlement neither the number of service supply points nor the highest level of service provided should be taken into account, but rather the range of goods and services as provided. Second, the provision of services obviously implies that they are used. Especially in the case of higher level services, not only the inhabitants of the settlement itself, but also inhabitants of the surrounding areas will be using them. The importance of a settlement for the provision of services to its surrounding area determines its centrality.

Services and Settlement Hierarchy
Large differences exist between services. As a consequence, the allocation of weights is required to establish the service level of settlements. Such allocation, however, is not an easy task since a number of arbitrary decisions are involved. First, how to weigh various services within one sector or group of services? For instance, what weight to allocate to a hospital and what to a primary health centre? Second, how to value individual units within one category? Should the quality of a service be taken into consideration? Should, for example, the presence or absence of qualified personnel be taken into account? Should accessibility matters be included - for instance, by considering opening times? Third, what about the different sectors

or groups of services? What is more important, credit supply points or health centres?

For our analysis of the pool of data, we have opted for an approach whereby each service obtains a score of `1' if present and a score of `0' when absent. If various levels can be distinguished within a certain sector or group of services, a simple weighing system has been applied, whereby for each subsequent service level one point is added to its score. Consequently, the hierarchical level of all settlements in Bantul district is based on scores as given for three groups of services, namely community services (0-13 points), production related services (0-12 points) and commercial services (0-40 points). (Details for the scores for these respective groups and the total score per settlement are included in annex 9.1). The relatively high number of commercial services influences the final score to a larger extent than the other two groups. Since most commercial services are more frequently used than the other types, such is not considered a distorting factor.

On the basis of the scores as obtained, the 75 centres in Bantul district have been grouped into five categories. As expected, Bantul town is the settlement with the highest score by far and forms a `group' by itself. Following ESCAP terminology (ESCAP 1979; DHV 1985)) this (level 1) settlement is referred to as 'regional city'. The second group, comprising five (level 2) settlements with scores between 39 and 45 points, has been labelled 'district towns'. The third group consists of ten (level 3) settlements with scores between 30 and 35 points; these have been categorized as `locality towns'. The remaining settlements are `rural villages'. As the scores for the villages in this group vary from a low of 4 points to a high of 29 points, two types of `rural villages' have been distinguished: the `high level rural villages' with (level 4) settlements with a score between 20 and 30, and the `low level rural villages' (level 5) settlements with a score below 20.

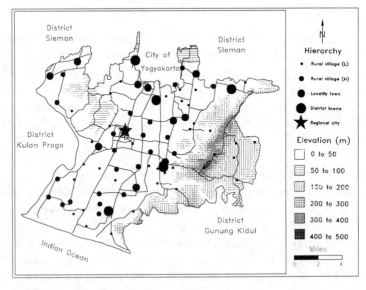

Figure 9.3: A hierarchy of settlements on the basis of all services
Source: Field research data

Figure 9.3 displays the hierarchical level for each settlement and in table 9.1 the subdistricts are listed according to the score of the highest level settlement. Here, a considerable differentiation can be observed. The service level in the settlements of Dlingo and Pajangan (the subdistricts with the most accidented topography), for example, is the lowest by far. Although two of the four settlements with the highest total score (Donotirto and Karangtalun) are located in the southeastern part of the district, the general level of services is mostly higher in the settlements in the northern and central parts in comparison to the southern part. From the eight highest scoring settlements, five are located in the northern zone. In contrast to this, from the nine lowest scoring settlements, six are located in the south.

Table 9.1: Settlements with the highest service level per subdistrict Bantul district, 1990

Subdistrict	Settlement*	Level	Subdistrict	Settlement	Level
Bantul	*Bantul*	1	Pandak	*Wijirejo*	3
Sewon	*Bangunharjo*	2	Pleret	*Pleret*	3
Kretek	*Donotirto*	2	Srandakan	*Trimurti*	4
Imogiri	*Karangtalun*	2	Bambanglipuro	Sumbermulyo	4
Piyungan	*Srimulyo*	2	Sanden	Gadingsari	4
Kasihan	Ngestiharjo	2	Jetis	Patalan	4
Pundong	*Srihardono*	3	Dlingo	Terong	5
Sedayu	Argosari	3	Pajangan	*Sendangsari*	5
Banguntapan	Baturetno	3			

Source: Field research data, 1990
* Names of subdistrict capitals are printed in italics

In the next two sections, we will determine the centrality of each of the settlements and subsequently analyze the extent to which the hierarchy of settlements as arrived at coincides with the centrality of those settlements.

Interaction and Settlement Centrality
Our data on interaction show that destinations inside Bantul district are by far the most important (table 9.2 - For details on the methodology see Note 1).

Three out of four interactions have a destination inside the district. Outside the district, Yogyakarta is the most important destination: one out of every five trips is aimed at this city. Neighbouring districts appear to play an unimportant role as regards the provision of services. Some who live near the district boundaries, cross the district border for certain services. However, the influence on the pattern as a whole is negligible; less than 3 percent of the trips are directed towards neighbouring districts.

The centrality of the various district settlements has to be assessed in conjunction to the role of Yogyakarta as a provider of services. This regional metropolis appears to be the only destination that is reported by the inhabitants of all settlements.

Table 9.2: Interactions from Bantul district according to main destination Bantul district, 1991, absolute figures and percentages

Destination	Interactions	
	Absolute	Percentages
District Bantul	21,912	75.9
Yogyakarta	6,111	21.2
District Sleman	485	1.7
District Kulon Progo	207	0.7
District Gunung Kidul	133	0.5
Total	28,848	100

Source: Field research data, 1991

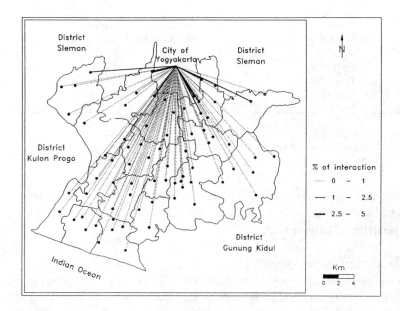

Figure 9.4: Interactions with the City of Yogyakarta as destination
Source: Field research data, 1991

Especially the people living in the peri-urban subdistricts are more focused on the city than on the district capital (figure 9.4). For the analysis of the centrality of settlements within Bantul district, a classification similar to the one arrived at earlier is used. Settlements are subdivided in five groups on the basis of their interaction pattern. Two criteria are used: the total number of incoming interactions and the relative importance of interactions with households outside the subdistrict of which they form a part. `Incoming interaction' is used for all interactions that end in the settlement concerned. Interactions may start either inside or outside the centre.

The rural villages are classified on the basis of the interaction with areas outside the own subdistrict. The classification of the other three groups is based on the absolute number of incoming interactions. The scores for the various settlements are presented in annex 9.2, a summary is presented in table 9.3. It appears that, except for the two types of rural villages, the interaction patterns are quite different. A decreasing level of centrality coincides with a decreasing number of incoming interactions, a decreasing service function for surrounding subdistricts and an increasing importance of internal interactions.

Table 9.3: Incoming interaction patterns for the five types of settlements. Absolute figures and percentages, Bantul district, 1991

Type of settlement	Average incoming interactions	Incoming interactions (percentages)			
		Internal	From own subdistrict	From other subdistrict	Total
Regional town	3725	11	21	68	100
District town	868	19	47	33	100
Locality town	243	54	26	20	100
High level rural villages	196	62	34	4	100
Low level rural villages	182	59	41	0	100

Source: Field research data, 1991

Table 9.4: Settlement with the highest centrality level per subdistrict, Bantul District, 1991

Subdistrict	Settlement*	Level	Subdistrict	Settlement*	Level
Bantul	*Bantul*	1	Kretek	Tirtomulyo	3
Imogiri	Imogiri	2	Sanden	Srigading	3
Pleret	*Pleret*	2	Jetis	Patalan	3
Piyungan	*Srimulyo*	2	Dlingo	Mangunan	4
Bambanglipuro	Sidomulyo	2	Pundong	Panjangrejo	4
Pandak	Wijirejo	2	Kasihan	Tamantirto	4
Banguntapan	Banguntapan	3	Pajangan	*Sendangsari*	4
Srandakan	Trimurti	3	Sedayu	Argosari	4
Sewon	Panggungharjo	3			

Source: Field research data, 1991
* Names of subdistrict capitals are printed in italics

In table 9.4, the settlement with the highest level of centrality is given for each of the 17 subdistricts. It is interesting to note that in only four of the subdistricts this concerns the subdistrict capital. Furthermore, the subdistricts bordering Yogyakarta do not have any settlement that is classified as `district town' or higher. Five subdistricts have no settlement with an interaction pattern that classifies them as higher than `rural village'. Four of these are located in relatively accidented areas.

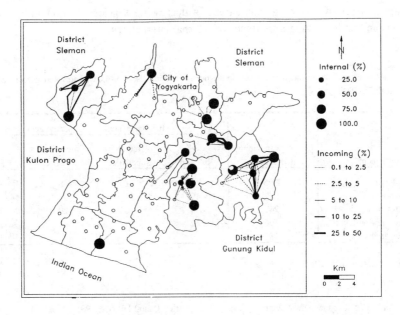

Figure 9.5: Interactions with `low level' rural villages as destination
Source: Field research data, 1990

`Low-level' rural villages
The bottom end of the hierarchy is constituted by those settlements that do not have any incoming interactions from outside the own subdistrict. This group consists of 19 settlements, two of which are Ngestiharjo (Kasihan). The latter settlement is even one of the largest in the district. Due to its location in the urban fringe of Yogyakarta, this settlement most likely experiences a fierce competition from its surroundings.

The spatial distribution of the lowest level settlements (figure 9.5), shows an interesting pattern. First, all but one of the settlements in the eastern-most (Dlingo) and western-most (Sedayu) subdistrict are in this group. Also Imogiri and the eastern part of Pleret show a relatively large incidence of this type of settlements. This is another indication that the level of services as provided in the uplands is very low. In contrast, nine subdistricts, mainly located in the southern and central areas, do not have any settlements that belong to this group.

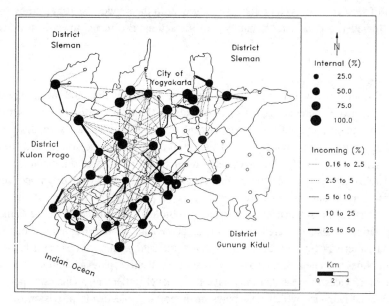

Figure 9.6: Interactions with `high level' rural villages as destination
Source: Field research data, 1991

Second, many of the settlements have only (a) very thin interaction line(s) attached, which indicates that most of the interactions are internal. For most of the settlements more than 50 percent of the interaction, and for half of them even 90 percent or more, originates in the own settlement. This limited service area may be seen as characteristic for low level service centres. In general, they have a relatively low number of incoming and a high number of outgoing interactions.

`High-level' rural villages
The 37 settlements belonging to the group `high-level' rural villages are depicted in figure 9.6. These rural villages receive visitors from outside the subdistrict, but not more than 10 percent of the total interaction. Seven of the settlements in this group are subdistrict capitals, most of which display a relatively high amount of interactions originating outside the own settlement, but within the own subdistrict. Five subdistricts (Dlingo, Sedayu, Kasihan, Pajangan and Pundong) do not have any settlement with a centrality that exceeds the level of a rural village. These five subdistricts are located in the eastern and western part of the district.

The interaction pattern for this group of rural villages is more complex than for the previous one. Although a large number of interactions takes place within the subdistrict and even within the settlement itself, there is also a fair amount of interactions from neighbouring, or even more distant, subdistricts. This implies that people are willing to travel longer distances to make use of goods or services as provided in those settlements. The services provided by this level of service centres, therefore, is additional to the services as provided by lower level centres. However, a considerable number of interactions is directed towards still higher level centres.

Locality towns
The group of locality towns comprises 13 settlements with less than 500 incoming interactions of which at least 10 percent originates from outside the own subdistrict. The settlements in this group are mainly located in the central zone of the district and are virtually absent in the eastern and western areas (figure 9.7). This group can be clearly distinguished from the two groups described previously. The average number of interactions to settlements in this group is considerably larger than for the earlier groups. Moreover, the relative importance of the settlements for its surrounding areas is larger; on average 20 percent of the interaction originates outside the subdistrict. The relatively limited importance of the settlement for its own population is also reflected in the lower percentage of internal interactions. From the interaction pattern in figure 9.5, it shows that some settlements have several relatively thick interaction lines attached. This indicates that these centres play an important role for the provision of services in areas that only have lower level settlements. This especially applies to Srimartani in Piyungan, which thus plays an important role in the provision of services to inhabitants of the adjacent subdistrict Dlingo.

District towns
This group comprises five settlements only. These tend to provide services especially to the population in the eastern part of the district (figure 9.8).

The `catchment area' of the settlements is fairly large and covers almost all settlements in neighbouring subdistricts and even many settlements that are located further away.
In table 2, the difference between this group of settlements and the lower level settlements becomes clear. For the lower level settlements, on average, at least 50 percent of the
interaction is of an internal nature. For this group, however, only one in five interactions has a local origin. Also the absolute number of incoming interactions is much higher for this group of settlements than for the lower level settlements. Especially Imogiri appears to play an important role in the provision of services to its surrounding area. The absolute number of interactions for this settlement is two to three times higher than for the other settlements in the group and only 12 percent originates in the settlement itself. Next to Bantul town, Imogiri is the second most important service centre in the district.

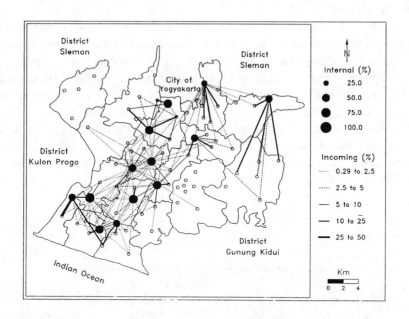

Figure 9.7: Interactions with locality towns as destination
Source: Field research data, 1991

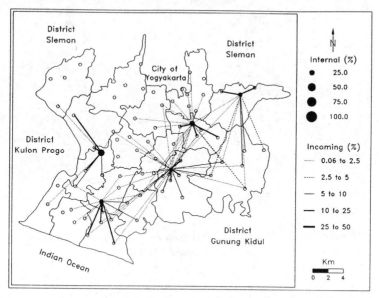

Figure 9.8: Interactions with district towns as destination
Source: Field research data, 1991

Regional Town Bantul

Bantul town is a category in itself. The number of incoming interactions is much higher than for any of the other settlements. The score is partly the result of the high level of services provided (figure 9.9). It is also caused by the large variety of services. Its important role as a regional centre can be clearly distilled from the number of interactions and the size of its catchment area. People from almost all settlements visit it for the provision of services. Main exceptions in this regard are some settlements in the peri-urban area and most of the settlements in Dlingo. The first group of settlements is oriented towards Yogyakarta. Inhabitants from subdistrict Dlingo, on the other hand, are more focused on Imogiri and Srimulyo (in Piyungan). Due to the accidented terrain these two places and Yogyakarta are easier to reach than the district capital Bantul.

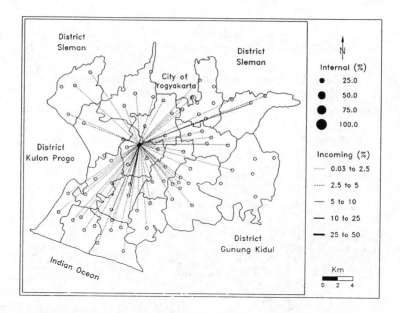

Figure 9.9: Interactions with regional town Bantul as destination
Source: Field research data, 1991

Hierarchy and Centrality Combined

So far, analyses have been made of the hierarchy and the centrality of settlements, respectively. Now we will attempt to establish whether settlements with a higher hierarchical level also have a higher centrality level. In other words, we will assess whether settlements with a larger range of goods and services and/or a higher level of goods and services are more frequently visited than settlements with a limited number of services of a lower level. Table 9.5 presents all the 75 settlements in Bantul district by level of hierarchy and level of centrality. Three different groups are distinguished. The first group, the ones on the diagonal, are the settlements that occupy the same level in the two classifications. From the figures on the diagonal it can be concluded that roughly one-third of the settlements in Bantul district has a centrality level that corresponds with their hierarchical level. The second group, adjacent to the diagonal, comprises the settlements that have a one class difference in the two classifications. This group contains roughly half of the settlements in Bantul district.

Table 9.5: Settlements in Bantul district by hierarchy and centrality

Centrality	Hierarchy					Total
	Regional town	District town	Locality town	Rural village (H)	Rural village (L)	
Regional town	1					1
District town		1	3	1		5
Locality town			2	6	5	13
Rural village (H)		2	4	12	19	37
Rural village (L)		2	1	4	12	19
Total	1	5	10	23	36	75

Source: Field Research Data, 1990-1991

Given the method of classification and the sometimes small differences between settlements in different groups, some leniency is required when comparing the two classifications. This leniency is exercised by qualifying the two classifications in this second group as matching. This leaves only few settlements for the third group that have a level of centrality that does not match with the hierarchical level. Six settlements have a centrality that is considerably higher than could be expected on the basis of their hierarchical level, while five settlements have a centrality that is considerably lower than could be expected. The first group of settlements consists of Sidomulyo and Mulyodadi (Bambanglipuro); Srigading (Sanden); Caturharjo (Pandak); Trirengo (Bantul) and Srimartani (Piyungan). The first four settlements are all situated in the south-western part of the district, an area with a limited number of higher level settlements. The second group contains the settlements Bangunharjo (Sewon); Ngestiharjo (Kasihan); Karangtalun (Imogiri); Donotirto (Kretek) and Argomulyo (Sedayu). The first three in this group are situated close to Yogyakarta and the centre of Imogiri.

Obviously, the service sectors of these three settlements face a fierce competition from these important centres.

Case Study: the Use of Community Services

Attention has been paid to the provision of social and commercial services as provided in Bantul district. Although both types of services are important for assessment of the hierarchy of a settlement and contribute to its level of centrality, the social services play a dominant role in the improvement of the living conditions of the rural population. Since the establishment of the New Order Government in Indonesia, three decades ago, an impressive number of facilities has been set up throughout the country. Especially in densely populated Central Java and the Special Province of Yogyakarta such has resulted in ease of access of households to basic services in the important fields of health and education (cf. Booth and Damanik, 1989; Huisman, 1994).

However, information about the actual use pattern of these services and the factors which determine this use pattern, is scarce. Below we will attempt to analyze these use patterns in education and health by rural households in our study area.

Patterns of Use

In the provision of formal education, a distinction can be made between primary schools (*SD*), junior high schools (*SMP*) and senior high schools (*SMA*). Together these form the backbone of the educational system. As a consequence of government policy, most educational facilities are distributed fairly even over the area. Without exception, villagers in the relevant age brackets are able to attend primary school in their `own' settlement. Not less than 80 percent of the lowest administrative level also provide for the first years of secondary education in junior high schools. Even the number of senior high schools is considerable; almost half of the villages has one or more senior high school. These senior high schools are mainly concentrated in the central, irrigated zone. The more accidented areas are less well-endowed in this respect, obviously a consequence of lower population density and relative accessibility limitations.

The health sector comprises of five different types of institutions, viz. hospitals, village health clinics, auxiliaries, mobile clinics and health posts. One private (catholic) hospital and one government hospital, with capacities of 50 and 90 beds respectively are present. The private hospital is of the lowest class in the Indonesian system (D). General practitioners form the medical staff; specialists are not available. The government hospital in Bantul town is of a higher category (C) and has five specialists. Both hospitals are responsible for curative care; preventive care is limited.

There are 21 village health clinics (*puskesmas*). Usually these are situated in the administrative capital of the subdistricts. Both curative and preventive care, including some health-related education is provided. The centres are headed by a medical doctor and often some trained para-medical staff is present. However, a shortage of adequately trained medical personnel may hamper an optimal functioning. In most cases, not all basic tasks as assigned to the clinics can be carried out, while opening times have to be limited to a few days per week.

The auxiliary clinics (*puskesmas pembantu*) are present in almost every village. Usually, they open only a few times per week. The auxiliaries are to provide simple curative care, mother/child care and family planning services. Medical doctors visit the centres, one or two health workers are responsible for the preparation of the set-up as required by the doctor on duty. On paper, every *puskesmas* is to be equipped with at least one mobile clinic (*puskesmas keliling*). This to enable visits to the more remote areas on a regular basis. Such, however, has not yet been realized in Bantul: only a couple of units are available. Due to the frequent use of the cars for other purposes, their effectiveness in rendering health services is questionable.

At the bottom end of the health services are the health posts (*posyandu*). These are staffed by villagers who have obtained basic training in general health care. Apart from first aid related activities, the staff is responsible for dissemination of information on the importance of hygiene in the household and other aspects of health-related ways of living. The *posyandu* thus fulfill limited roles only and have, consequently, not been included in our analysis.

In addition to these health services, private practitioners are present in the area. Most of these physicians can be found in the larger service centres of the lowlands zone. The location of services as actually used by the households is shown in figure 9.10.

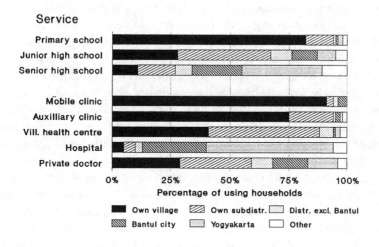

Figure 9.10: Location of education and health services as actually used by households (percentages)
Source: Field research data, 1991

A distinction has been made between services used in the own village, those elsewhere in the own subdistrict, those inside Bantul district but outside the town of Bantul; those in Bantul town; those in the city of Yogyakarta and those in other locations. It clearly shows that a large variety exists in the location of services as used. Apparently, the availability of a service in a village does not necessarily imply that this service is used by all households in that settlement. People may prefer to use a facility located outside the village even if such implies the bridging of considerable distances.

Although all villages in our study area have at least one primary school, from one out of every five households with children of primary school age, a school outside the own village is attended. According to information obtained, the actual walking distance is the most important factor to explain this phenomenon.

Higher level educational facilities are also relatively abundant in our study area. In this case, however, other factors than mere distance are playing a role in the decision as to which school a child is being sent. Although some 80 percent of the settlements provide junior high school education, only one third of the households sends children to a school within the own settlement. Almost four out of ten households sends children to a junior high school in the own subdistrict. The remainder uses services as located relatively far away from the place of residence. In the case of senior high schools, this pattern is even more clearly visible. One third of the households sends children to Yogyakarta, an additional one fifths to Bantul city. The own village and the own subdistrict, thus, are relatively unimportant for this level of education. Apparently, as the children grow older, the quality of the education or the variety of educational opportunities, play increasingly important roles.

The visits to health services display a pattern that is largely comparable to the one as shown by the educational services. The relative importance of the own settlement decreases with an increasing level of service provided. It is evident from the graph that up to the level of the village health care centre, the services are sufficiently provided within the own subdistrict. With an increasing level of service, the relative importance of the own village decreases, but this is counterbalanced by an increasing importance of services outside the village but within the own subdistrict. In total, some 90 percent of all households visiting a clinic or village health service, do not go outside the own subdistrict for this purpose.

For visits to a private doctor, the picture is somewhat different. Although the services provided within the own subdistrict are the ones most frequently selected, doctors located outside the own subdistrict are visited by four out of every ten households. Both Bantul town and Yogyakarta account for about one sixth of the households' visits with this purpose.

The visits to hospitals is different The class D hospital, located in Bambanglipuro, is hardly ever selected. Most important is Yogyakarta; over half of the visits is directed at hospitals in the regional metropolis, the C class hospital in Bantul town attracts a quarter of all visits for this purpose.

Factors Influencing the Use of Community Services

Emphasis has been put on where people go to visit a health or education service. Before addressing social differentiation in the use of social services, a few remarks are in order here on the relative importance of these services. The use of both education and health services varies considerably (figure 9.11). It needs stressing that the frequency of visits has not been included; households have been asked only to indicate whether or not a service is being used by any member of the household. The enrolment in primary education is more common than enrolment in higher levels: 44 percent of all households has one or more children attending primary education, as compared to one quarter for junior high school and only 13 percent for senior high school.

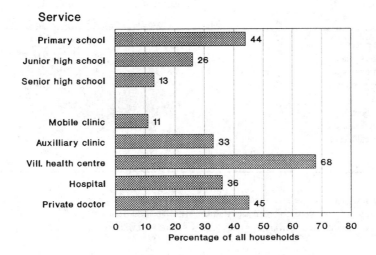

Figure 9.11: Use of education and health services in Bantul District (percentages)
Source: Field research data, 1991

The use of health services appears to be fairly common throughout the study area. By far the most commonly visited facility is the village health centre. Seven out of ten households make use of these centres. Less common, but still visited by one out of every three households are the auxiliary centre and the hospital. The mobile clinics play a relatively unimportant role in the provision of health services. Also the private health services are important. Private doctors are more often consulted than auxiliary clinics or hospitals.

Below, the analysis focuses on the actual use of services in relation to the socio-economic characteristics of the (head of the) household. The use of services is related to the agro-physical zone in which the household is located, the educational level of the head of household, the employment status of the head of household and finally, the household income.

Agro-physical zone
The use of social services does not present a very clear picture when related to the agro-physical zone in which the households are living (table 9.6). The accessibility of schools is relatively good for all zones. This may explain why the percentages for the various zones deviate only slightly from the figures for the study area as a whole.

Table 9.6: Use of services by agro-physical zone (percentages)

Type of service	Percentage of households per zone and per service.				
	Zone 1	Zone 2	Zone 3	Zone 4	Total
Education services					
Elementary school	51	53	38	41	44
Junior high school	28	34	24	23	26
Senior high school	8	8	16	16	13
Health services					
Mobile clinic	3	13	11	14	11
Auxiliary health clinic	14	56	39	32	33
Village health centre	81	62	65	65	68
Hospital	30	35	34	40	36
Private doctor	50	34	48	45	45

Source: Field research data, 1991
Zone 1 = peri-urban area; Zone 2 = rural area with less than 10 percent irrigation;
Zone 3 = rural area with 10-35 percent irrigation; Zone 4 = rural area with 36 percent and more irrigation

Whereas the mobile clinics are insignificant for the population in the peri-urban zone, the village health centre is relatively popular in the area. Hospitals are less frequently selected in this zone in comparison to the other zones. Village health centres in this zone generally have better qualified staff and more convenient opening hours than those elsewhere.

The relative popularity of the auxiliary health clinic in zone 2 can be explained by the topography of the area: some parts are relatively poorly accessible and people prefer to visit the most conveniently located health centre, rather than traveling all the way to better equipped higher level centres.

Level of education of the head of the household
The enrolment rate in primary education is almost complete; enrolment rates for the higher levels of education are subtantially lower. The decreasing rates with increasing levels shows clearly from our data. Surprisingly, however, no clear relationship can be established in this

regard with the level of education of the head of the household (table 9.7). Slightly more children from heads of households with more than primary education attend higher education. It cannot be concluded from these data, however, that children from higher educated parents are more inclined to pursue higher level education.

On the other hand, the visit of health services appears to be closely related to the level of education of the head of the household. Heads of household with more than primary school show preference for visits to a private doctor or a hospital above the visit to a health clinic, whenever medical advice is required. The opposite pattern can be observed in the groups with lower levels of education. Apparently the quality of care, or at least the reputation of supply points in this respect, is higher in the first group of services as distinguished.

Table 9.7: Use of services by educational level of the head of the household (percentages)

Type of Service	Percentage of households per educational level and per service			
	no formal education	primary school	more than primary school	Total
Education services				
Elementary school	42	46	41	44
Junior high school	15	28	27	26
Senior high school	10	12	16	13
Health services				
Mobile clinic	15	11	9	11
Auxiliary health clinic	47	37	21	33
Village health centre	83	70	58	68
Hospital	34	33	41	36
Private doctor	37	42	54	45

Source: Field research data, 1991

Status in Employment

For the influence of status in employment upon the use of social services no common denominators for the tracing of a pattern can be distinguished (table 8). Self-employed persons do not present a picture that is significantly different from the one of employees or civil servants. In addition it shows that no clear distinction can be established between `urban based' and `rural based' occupations.

Table 9.8: Use of services by status of employment of the head of household (percentages)

Type of Service	Percentage of households per employment status and per service						
	Unemployed	Self-employed		Employees	Civil Service	Other	Total
		Farmers	Other				
Education services							
Elementary school	31	38	49	56	36	50	44
Junior high school	18	23	31	29	28	21	26
Senior high school	6	16	12	9	17	12	13
Health services							
Mobile clinic	3	12	10	11	11	9	11
Auxiliary health clinic	44	38	28	41	19	27	33
Village health centre	76	70	70	76	53	68	68
Hospital	47	33	36	28	49	36	36
Private doctor	40	40	50	46	51	47	45

Source: Field research data, 1991

Household Income

The most decisive factor influencing the use of social services is the household income (table 9). Enrolment in primary education decreases slowly when household incomes increase. As regards the enrolment in senior high schools, a clear opposite of this pattern becomes visible. Households with children attending primary school generally are young families, who have relatively recently started to generate an income.

Moreover, since the children are attending school they are not (yet) in the position to contribute to the household income. On the other hand, the older children at senior high school age generally are from more mature households. In other words, household income appears to be closely related to the stage in the family life cycle. From the enrolment in junior high schools no pattern shows.

Regarding the health services, the relationship is more straight forward. With an increasing income, the relative use of lower level health facilities decreases; the use of higher level facilities increases. Especially the visits to private practitioners and to hospitals are much more expensive than those to health facilities such as clinics.

Table 9.9: Use of services household income (percentages)

Type of Service	Percentage of households per income group and type of service use						(income *Rp 1000 per month)	
	0-39	40-79	80-119	120-159	160-199	200-239	240+	Total
Education services								
elementary school	49	50	44	48	40	34	33	44
junior high school	22	22	29	30	26	18	28	26
senior high school	11	14	10	15	11	16	16	13
Health services								
mobile clinic	15	12	10	12	8	15	7	11
auxiliary health clinic	61	41	32	27	28	19	15	33
village health centre	81	76	73	67	61	50	53	68
hospital	31	32	35	35	40	46	44	36
private doctor	38	44	42	44	50	49	54	45

Source: Field research data, 1991

Conclusion

A wealth of empirical data on both supply and actual use of rural services has allowed us to analyze the role of the small rural centres in Bantul district, as situated just south of Yogyakarta, vis-à-vis the role of the regional metropolis in this respect. Special emphasis has been put on community services, and the results have given insight in the supply and use patterns pertaining to health and education. The influence of the regional metropolis is clearly noticeable throughout Bantul: Yogyakarta is the only destination that is visited by inhabitants from every settlement in the district and roughly one fifth of all interactions are aimed at this city. However, the extent and nature appear to vary considerably within the district due to contextual factors, but also within smaller areas due to socio-economic differentiation. Two main factors influence the supply of services in the area. An important contextual factor is the high overall population density which has made possible that in many subdistricts threshold levels are reached for numerous social and commercial functions. In addition, the government policy pursued constitutes an important influential aspect; for a range of social services, especially at the lower level, a blanket approach has been followed, which implies that these services are widely available. The provision of a good or service, however, does not automatically imply its use. The findings indicate that certain groups of users prefer services provided in Yogyakarta or Bantul Town, although the service is also provided at nearby destinations. This especially applies to those who have the ability to travel, i.e. the younger, higher educated and more affluent rural strata. Especially higher level services are frequented in Yogyakarta and both the use and frequency of use of those services appear to be strongly related to level of income. In other words, despite an efficient transportation infrastructure - which has resulted in relatively low transportation costs in most areas - and despite a long since mobile population, distance still plays an important role in the patterns of service use. As observed in the introduction, economic growth in the area has been strong for more than two and a half decades. As a consequence, the average rural households' income has risen considerably. This implies that, increasingly, services in the city have become accessible for larger segments of the population, which, in its turn may have occurred at the detriment of services provided by the rural service centres. Due to an increasing flow to the city, for an expanding range of goods and services, threshold levels may no longer be reached in the settlements at the lowest levels and, consequently, these supply points may disappear. Such disappearing of goods and services will start with commercial services, because their presence is purely based on economic considerations. Eventually, however, also community services may have to follow suit. At first probably through a decline in quality, but later also through a decline in quantity. Because the fruits of growth are not distributed evenly over the population, this process may imply that, progressively, poorer households may start to face difficulties in obtaining certain types of required goods or services from places nearby their residence. As a consequence, they may be forced to refrain from use, or have to allocate a higher share of their marginal incomes to obtain services in Bantul Town or Yogyakarta. To ensure that the quality of life and the access to essential services by the lower income groups in rural Bantul will not deteriorate, government policy makers should take account of these processes. To avert socially and politically highly undesirable forms of polarization in development, they will have to closely monitor the distribution aspects of the supply and use of services.

Notes

1. In 1990 and 1991 extensive field work has been carried out to build a database on the services delivery system in Bantul and the use made of these services. In each of the district's 75 settlements a stocktaking exercise of all services and their basic characteristics has been executed. Subsequently, 30 households in each settlement were randomly selected and interviewed to record their socio-economic characteristics and to inventorize their use of services. The frequency of visits paid by the households has not been included: members were asked to indicate whether or not a service had been used by any member of the household in the previous twelve months. In case more than one settlement had been visited by household members for a certain purpose, the most frequently used destination was recorded. This comprehensive data collection has resulted in a total of 28,848 interactions.
Further details are presented in Annex 9.1 regarding the allocation of weights to the various groups of services identified and the total score as arrived at per settlement. Besides, both absolute figures and percentages are presented in Annex 9.2 regarding the incoming interactions for all settlements.
2. So far, data collection has taken place only once and, therefore, similar exercises are necessary to analyze the dynamics in the provision and use of services.

CHAPTER TEN
The Small Town Reconsidered

Milan Titus

This final chapter attempts to analyse the preceding case-studies in their wider functional-spatial and political-economic contexts with the purpose of testing the theoretical assumptions discussed in chapter 2. In doing so, we are directly confronting the problem of the limited range and validity of our case-studies. To what extent is it possible to draw valid general conclusions on the structure and role of small towns from a few isolated and very different case-studies? On the other hand, we should not *a priori* hide ourselves behind a facade of easy arguments and refrain from this intellectual responsibility. Therefore, we have tried to solve this dilemma by comparing and testing the conclusions from our case-studies with those offered by other cases and more general publications on small and intermediate urban centres (SIUCs).

This final chapter will deal with the following functional aspects of small towns in general and of our case-studies in particular: 1) population characteristics and dynamics; 2) production structure and production relations; 3) internal and external linkages: 4) public sector and public services; 5) labour market and employment; 6) regional development performance, and finally; 7) the main theoretical and policy implications. In each of these sections special attention will be paid to the role of government policies and interventions as both the national and regional governments still are the dominant actors in regional development and planning in most developing countries, where small towns often are assumed to fulfil a central role in transmitting and implementing these policies at the local level (cf. Rondinelli 1990; Gaile 1992).

The Small Town Population Characteristics and Dynamics

Although there is now general agreement that there is no universal definition of small and intermediate urban centres according to population size, it is also commonly accepted that SIUCs occupy the lower end of the rank-size hierarchy and display a limited range of servicing functions for their immediate hinterland areas. This implies that there is at least a remote relationship between the size and growth of SIUCs and their functional characteristics (cf. Hardoy & Satterthwaite 1986, pp. 279 ff), as well as between their functional characteristics and the rural hinterland conditions (Gilbert & Gugler 1992, pp. 50-53). In principle, however, the first type of relationship applies only within the same urban system and is not confirmed by our case-studies from a cross-national perspective. Relatively large SIUCs, such as Mopti (74,000) or Purbalingga (30,000), do not appear to be economically more important or dynamic than smaller centres like Wonosobo (26,000), Ciudad Quesada (13,000) or even Nhlangano (4,000). This is because their size tells very little about the economic potential of their respective hinterlands, nor about the types of relationships with these hinterlands. The latter aspect, for example, severely restrains the development of Purbalingga which suffers from competition posed by the nearby higher order centre of Purwokerto (150,000). Similarly, Mopti's development is severely hampered by the debilitating impacts of periodical droughts in its hinterland area. On the other hand, smaller centres may enjoy a much more favourable position as semi-monopolistic centres in a dynamic hinterland setting driven by processes of agricultural colonisation and commercialisation. For these obvious reasons there is no simple and direct relationship between population size and growth among the various types of SIUCs.

More convincing evidence has been found on the relationship between small town population growth and the functional importance of these centres if these parameters are related to the economic hinterland conditions. In Mopti, for example, periods of low and moderate population growth alternate with fluctuations in rural hinterland conditions and food supply. Even during periods of severe drought and food shortage, the town keeps growing slowly in spite of its nearly stagnant urban economy. This is mainly due to the town's central distribution role for food aid and for some basic public and private services to the region, thereby showing that its functional dominance as a regional service centre is a more or less permanent feature and a decisive factor in regional development.

The two other larger towns, Nueva Casas Grande (49,000) and Manzini (43,000) comply more with the assumption of a direct relationship between economic development and urban population growth. In the case of Nueva Casas Grande this growth, however, is primarily determined by externally induced processes of economic development through the establishment of so-called *maquiladora* enterprises and not so much through rural hinterland development. In the case of Manzini, the second urban centre of Swaziland, there is a much clearer relationship with rural hinterland development based on agricultural surplus production. But here too, strong external impetuses to its growth and functional position were given through the establishment of some major administrative and public services, as well as through foreign private investments in trade and processing industries.

Smaller centres usually have more problems in maintaining high population growth rates as these presume economic development through improved infrastructural facilities and changing patterns of production and consumption. These in turn may easily invite activities and services from higher order centres to invade the small town's hinterland and erode its servicing functions. This has clearly been the case in the Serayu Valley region in Central Java where Purbalingga shows a much lower growth rate than Wonosobo, although both towns belong to the same rank-size order and perform similar administrative functions as district capitals. Moreover, both hinterlands are densely populated rural areas with commercial types of foodcrop farming, but in the case of Wonosobo there is no competing higher order centre tapping the hinterland's surpluses and limiting its size.

A similar situation is found in Bantul District (Special Province of Yogyakarta) where both the district capital of the same name and its lower order rural service centres are suffering from the competition effects of the nearby provincial capital, Yogyakarta, that offers a wide range of higher rank services and non-farm employment opportunities.

The smallest centres among our case-studies, Ciudad Quesada and Nhlangano, both show relatively high population growth rates. This is probably due to the fact that these more or less recently established centres are the only existing servicing centres of any importance in an extensive area. Moreover, their rural hinterland economies have already started to generate considerable surpluses through the settlement of increasing numbers of agricultural colonists and the switch to commercial types of farming.

The Small Town Production Structure and Production Relations

With respect to the production structure characteristics of SIUCs it is mostly assumed that the larger centres usually display a more diversified structure than the smaller ones (Davila & Satterthwaite 1987; Hardoy & Satterthwaite 1986, pp. 279-284). The main reason for this is that in larger urban centres formal or corporate-type activities have better development opportunities than in small centres, due to the relatively high threshold values for these types of activities in terms of population numbers, purchasing power and production facilities. This implies also that production relations in the larger centres are more strongly determined by modern capitalistic principles, such as the use of wage labour, formal credit and formal marketing channels by manager-operated enterprises, instead of the pre-capitalistic relations predominating in the owner-operated family firms and the independent petty producers in the smaller centres.

In economic-sectoral terms, SIUCs usually display a predominance of private and public service activities, whereas the secondary sector activities often remain underrepresented due to small hinterland surpluses and weak backward linkages, and/or the competition from more efficient industries in higher order centres or abroad. This unbalanced economic structure is also completely in line with the dominant servicing role of the small towns in the lower ranks of the urban hierarchy, as well as with their local-level administrative functions. The sectoral structure of the labour force in our case studies also shows clear concentrations in the servicing sector, and particularly in the (semi-) public services and the commercial services. Both these sub-sectors usually are by far the most important in terms of employment and income generation. The size and role of the usually less prominent secondary sector activities, however, may still vary considerably according to the following conditions:

- The opportunity to supply a local market with cheap artisanal products and processed food which serve the basic needs of the poorer sections of both the urban and the rural hinterland populations. This situation applies in particular to the surviving petty producers in Mopti, where very small rural surpluses and poverty conditions protect the local market from the competition of imported and more expensive and durable items. These traditional types of industry are mostly involved with the production of basketry,

pottery and wood and leather products. Most of these activities are of a *supply-push* type, created by a permanent labour surplus.
- The opportunity to process rural raw materials and surpluses into higher quality food products or building materials, and to produce more sophisticated consumer goods like clothing, leatherware or furniture for a wider (national) market. Although many of these activities are still of a *supply-push* type, they usually have better prospects in facing external competition, because of cheap labour and local skills involved in producing (semi-) quality ware. This situation is applicable to the Indonesian case-studies, where increasing pressure on the local labour markets is accompanied by increasing agrarian surpluses and slowly rising incomes, which drive ever more entrants of the labour market into non-farm activities.
- The opportunity to attract national and foreign capital investments to small towns which offer special advantages in exploiting local physical and human resources and enjoy locational advantages with respect to international markets. This usually leads to the establishment of so-called *company towns* in mining or plantation areas, and of so-called *maquiladora towns* in borderland areas with an abundant supply of cheap labour and a well-developed infrastructure. Manzini (Swaziland) and Nueva Casas Grande (Mexico) are more or less exemplary for this type of small town development.

This synopsis of the main characteristics and determinants of the small town economic structure makes clear that there is no direct and unambiguous relationship between their population size and the relative size of each of their economic sectors, nor between the size of the respective economic sectors as such in a sense that a large servicing sector is always supported by a dynamic and leading secondary sector. The sizeable tertiary sector in these towns, on the contrary, often reflects their unbalanced functional structure and lack of economic dynamism, pushing the majority of their labour force into tertiary refuge types of activities like street selling, market trade, local transport and domestic services.

As far as public services are concerned, these usually are more developed and organised on a formal basis due to strong governmental commitments. This applies in particular to public facilities in education, health, administration, and in energy supply and communications, which all belong to the central functions of regional service centres. Because of frequent over-staffing, weak performance and low salaries however, many of these public services show *refuge sector* characteristics as well.

When production structure characteristics of small towns are analysed in terms of predominantly capitalistic or non-capitalistic relations of production (see chapters 2 and 8), the corporate-type enterprises in our case-studies usually are strongly underrepresented vis-à-vis the petty-type activities, i.e. both in numbers and in employment opportunities. Among these corporate-type enterprises there are only a few large scale and capital-intensive industrial or trade companies that are also manager-operated. The majority of the corporate enterprises are owner-operated *family firms* like shops, and smaller whole-saling, trading and transport companies. Although these corporate-type enterprises certainly are most important in terms of gross regional product generation, most corporate sector employment opportunities are generated in the (semi-) public sector, including the local government services. Moreover, the money circulating in these towns and their hinterlands usually also stems mainly from public expenditures for salaries and investments in infrastructural projects. Consequently, corporate sector development in these towns still is very much dependent on government programmes and public spending (cf. Hardoy & Satterthwaite 1986; Gaile 1992). A major exception to this rule of course is offered by the so-called *company towns* where private sector investments and activities are clearly dominating the town's economy and the corporate sector usually is the dominant sector in terms of size, employment and production value.

Again, there are few indications that there is a direct relationship between a town's size and the size of its corporate sector economy. Although it is true that very small towns often lack the facilities, customers, purchasing power and qualified labour to generate sizeable corporate sector activities, this does not mean that the establishment of such activities is categorically blocked from these centres. As such the cases of Nhlangano, Manzini or Nueva Casas Grande tell instructive stories, proving that external development factors and rural hinterland conditions often are much more decisive. Mopti, the largest regional centre among our case-studies, offers strong evidence in a negative sense. Here several large scale state enterprises had been established in the 1960s with the purpose of processing rural surpluses and supplying the region with basic commodities. But soon after the collapse of the rural economy nearly all big enterprises went bankrupt due to their intrinsic lack of flexibility, whereas many petty enterprises were able to survive (Harts-Broekhuis & De Jong 1994, p. 274). Similar experiences were found in the Serayu Valley towns in Central Java as a consequence of structural adjustment policies in the mid-1980s (Steenbergen & De Jong 1987; Van der Wouden 1997, p. 96). Moreover, size and dynamics of the corporate sector economy here, were not so much related to their respective population sizes, but rather to specific historical conditions like the presence of a Chinese entrepreneurial middle class (Wonosobo, Purbalingga), as well as to their rural hinterland conditions and so-called *bypassing* effects from higher order centres.

The latter factor is especially important for the explanation of the weak economic structure of Banjarnegara town, that lost nearly all its main collecting trade functions. The impact of bypassing evidently is also playing an increasing role in sapping the economic structure of Ciudad Quesada (Huetar Norte). Initially the town's economy grew successfully by handling increasing agricultural surpluses (coffee, sugar) from the expanding colonisation areas in Huetar Norte. But with the implementation of structural adjustment policies and the subsequent switch to extensive cattle ranching for beef export, the town's position became gradually weaker as most cattle is exported directly to the world market via San José (Romein 1995, p. 135). Moreover, declining demand for agricultural inputs and low purchasing power among the *hacienda* workers, reduce the demand for urban goods and services, thereby further debilitating corporate sector growth in this town.

On the other hand, corporate sector development may be particularly dominant in the so-called company towns which have been established on the basis of foreign capital exploiting and processing local natural resources. The singular focus of these towns on serving export-oriented production of raw materials and industrial cash crops often makes them very vulnerable to the vicissitudes of the world market. The more so, since their distributive functions usually are confined to serving the local company staff and employees, while the local authorities and companies often try to deter the establishment of informal sector activities. However, in cases where these towns have been able to attract other functions, like administration and public services, or have been able to expand their services to rural producers in their hinterlands, their economic structure may become more diversified and stable. This has clearly been the case in the towns of Manzini and Nhlangano in Swaziland, which distinguish themselves favourably from the stagnant mining towns in for example Potosí District in Bolivia (cf. Evans 1989).

Finally, corporate sector development may take place more or less independently from local resources or hinterland conditions through massive investments from abroad and from domestic sources. This situation has been encountered in the Northern Mexican borderland area, where United States capital has started a spectacular process of industrialisation in many local small and intermediate centres. This has been achieved through the establishment of both foreign branch companies and local sub-contracting firms, attracted by conditions of cheap labour, infrastructural and tax facilities, and last but not least by the proximity to the United States market. The case of Nueva Casas Grande shows that this spectacular corporate sector development has also attracted many informal sector workers serving the needs of the increasing number of formal sector firms and employees. Processes of small town development and economic diversification, therefore, can no longer be analysed solely in terms of local and regional conditions.

Internal and External Linkages of the Small Town Economy

According to the more recent and accepted interdependent view on the production structure of Third World urban centres (Santos 1979; McGee 1981; Moser 1984; Roberts 1978, 1995), the functional relations between the formal and informal sector activities viz. the corporate and petty-type activities, can no longer be considered as being primarily of an antagonistic nature. Instead, their functional interrelatedness implies that they may coexist in mutual support, although this does not exclude exploitative relationships. Consequently, the expansion of the formal sector (i.e. corporate enterprises and government institutions) does not *per se* lead to the informal sector's disappearance, while in practice it even may engender its parallel growth. The main reason is that on the one hand informal sector activities tend to support formal sector growth by supplying cheap food, inputs and services, while on the other hand they take advantage of the increasing circulation of money and goods generated by this formal sector growth.

Indeed, little or no evidence was found in our case-studies in support of the classical view on formal-informal sector competition. In as far as there were any large scale, manufacturing enterprises in the small towns, these were mainly oriented towards the national and international markets, and seldom or hardly used local raw materials. This contrasts with the local small scale and intermediate size enterprises, which usually served local or regional markets and practically all utilised local raw materials (crops, minerals, wood, hides, etc.) as well as local suppliers and traders in the petty-type sector. The weak backward linkages of the larger manufacturing enterprises are also demonstrated by an obvious lack of sub-contracting relations with the smaller or petty-type enterprises. This applies especially to manufacturing industries established by foreign capital, and utilising cheap local labour, production facilities and international marketing opportunities, but operating in a semi-detached, footloose way in industrial enclaves like the so-called *maquiladora* centres in Northern Mexico.

Similar features of weak local integration, however, were also found among the larger *autochthonous* types of industry in the Central Javanese towns and in Mopti. Another remarkable aspect of these larger manufacturing enterprises is that they do not always occupy a stronger position in facing competition or market slumps than the small scale enterprises (Titus & Van der Wouden 1992). For reasons of dependency upon external capital, technology and (inter)national markets, and because of a lower degree of flexibility, they often are more vulnerable to the impacts of failing demand, budgetary cuts and high rates of inflation than many petty producers who serve the poorer and relatively protected segments of the local market. This applies in particular to the manufacturing of *traditional* goods like processed food, pottery or mats, which are not profitable to the corporate-type industries. In some cases - like in Purbalingga - petty producers even succeeded in securing stable markets outside their own region by producing car spare parts and kitchen utensils for specific segments of the urban market.

Similar signs of the viability of small scale industries producing more 'modern' types of consumer goods like garments, furniture and building materials, have been noted in the Egyptian town of Fayoum (Hoffman 1986). Most of these small scale manufacturing activities in small towns, however, can only meet local demand and fulfil local basic functions, due to their poor capital and technology base and their exclusive orientation towards a 'traditional' assortment of commodities for the poorer sections of the population.

In summary one may conclude, therefore, that neither the corporate nor the petty-type manufacturing activities in the small towns are able to play a strong basic role for the regional economy. On the contrary, the services sector in the small town economy seems to have developed much stronger local and regional ties. The relations between the commercial corporate-type and petty-type activities, usually are not only more direct, but also of a more symbiotic nature than those in the industrial sector. Wholesalers residing in the market centres, for example, often make use of the services of petty traders operating in the rural hinterland areas for collecting the various surplus products. This is often achieved through informal credit systems binding the small farmers and non-farm producers to specific urban traders, thus enhancing the surplus extraction from the hinterland areas. Tobacco trade in Wonosobo and cattle trade in Mopti are both illustrative cases for this type of relationship.

On the other hand, most traders in street selling, petty shops and petty restaurants buy part of their inputs and commodities from local corporate sector shops and commercial agents. Again, many of these products are supplied on credit with the provision that the petty trader pays after selling his product to the consumers, thus binding the trader to his supplier. Moreover, there seems to be little competition between the corporate and petty-type activities in distributive trade, as both types usually serve different market segments. Fiercer competition, however, is found among the petty traders themselves, as these are continually striving for a larger share in a limited market by offering special bargains, thus undercutting the average market prices. Consequently, value added in these small scale types of distribution trade usually is very small and petty trade is characterised by high turnover rates among marginal entrepreneurs. This however does not deny the existence of more successful types of petty activities in the small towns, like the technically more advanced types of repair services and motorised transport services.

Commercial services in small towns usually are not only put under pressure by internal forces like the overcrowding of the petty type sector, but also by external forces of competition from higher order centres. This latter type of competition may affect both the distributing and collecting trade functions of the small town. In distribution trade, it especially is the impact of corporate-type enterprises in larger cities, enjoying certain economies of scale and taking advantage of improved communications, that is increasingly undermining the small town's hinterland functions. Usually this is achieved by employing local trade agents and offering special credit facilities to their local customers. In addition improved transportation facilities enable the emerging rural and small town middle classes to buy their consumer durables more often in larger cities which offer their customers a wider choice.

In collecting trade the small town's functions often are undermined as well because urban wholesale traders increasingly go directly to rural producers in order to collect the growing surpluses resulting from technological innovations in agriculture. This development usually is welcomed by the farmers who may enjoy higher farm gate prices due to the short-circuiting of the trade flows. Thus, both the 'break of bulk' and the 'critical mass' collecting functions of the small towns are progressively being eroded at the regional level by better communications, advancing commercialization and increasing economies of scale.

The relevance of these processes for our case-studies has been clearly demonstrated in the changing rural-urban relations of small towns in Huetar Norte (Costa Rica) and in the Serayu Valley region in Central Java (cf. Romein 1995, pp. 84-89; Van der Wouden 1997, pp. 145-146). Nevertheless, many small town-based wholesale and petty traders have been able to keep control of at least part of the rural hinterland trade by carefully adapting their prices and credit facilities and by utilising the advantage of proximity to their producers and clients.

The Small Town Public Sector and Public Services

In several publications Rondinelli (1983, 1988, 1990) has pointed out the central role of the small town public sector in supplying its hinterland population with essential administrative, educational, health and extension services. The dominant position of the public sector in the small town economy usually is demonstrated by the relatively high share of qualified employment opportunities created by this sector, as well as by its direct economic impact through public expenditures, development programmes and licenses. Consequently, the local bureaucracy in these towns often appears to be more influential than the weakly developed entrepreneurial middle class, and sometimes their distinct social positions may even become completely blurred when civil servants exploit their privileged positions 'to go into business' like in the Central Javanese towns. Moreover, much of the money circulating in the small towns stems directly from salaries earned in the various public sector jobs.

On the other hand, the local government may skim off part of the rural hinterland surplus through local taxes and interest on public sector loans. Fragmentary insight into the balance of public expenditures and incomes at the regional and local levels in different countries like Indonesia and Mali however, suggests that this balance is in favour of the hinterland economies, if subsidies on food and inputs are included. This implies that the public sector's role as a pacemaker for the small town and rural hinterland economies is a very vulnerable one in case structural adjustment programmes force governments to cut down their public spending levels. Especially the small towns with top-heavy administrative and public servicing functions like Banjarnegara or the Malian towns tend to be severely hit. This contrasts with the so-called *company* or *maquiladora* towns, which have an important corporate-type sector producing for export markets that enables them to derive advantage from deregulation measures and compensate for possible slumps in domestic demand.

Because of the frequently dominant position of the public sector in the small town economy, one would expect their public servicing role to be particularly well developed. A closer look at our case-study towns however shows that there are considerable differences between the various centres. These differences are partly due to the types and quality of the urban facilities available, but even more to the intrinsic imbalances in rural-urban relations, prevailing rural relations of production and the accessibility of the services. Van Teeffelen's contribution on the functioning of rural service centres in Southern Mali for example, shows that the presence of urban public services does not guarantee their effective use by the intended target groups, who often lack the necessary purchasing power, knowledge and motivation.

This applies in particular to the utilisation of the more advanced types of public and private services, such as secondary schools, specialised health services, formal credit services, telecommunications services and mass-media services. If present at all, these services are mainly used by the local urban middle class and hardly by the rural hinterland population itself. The local market for those higher quality services often is further constrained by the fact that the more mobile urban and rural middle classes prefer to use the higher quality services in the nearest big town or city, thus by-passing their own service centre.

Moreover, even if the small town services are aimed at the poorer sections of the rural hinterland population, their quality often is ill-adapted to the needs and capacities of these target groups. Notorious examples of such incongruities are mentioned by Van Teeffelen with respect to the poorly performing (semi-) public services in the Malian towns in the fields of agricultural extension, rural development and marketing (boards) as well as in health care and education.

This means that the use of public servicing functions of small towns is as much determined by the quality of the services offered, as it is by factors of physical accessibility or local purchasing power. Considering that the main responsibility for the quality of public services usually rests with the national and regional governments, they are also the ones who should initiate any measures for improvement. The same conclusion applies to the observed detrimental impact of rural marketing boards on both the small town and the rural hinterland economies. The situation whereby parastatal boards supply credit and inputs, and buy food and cash crops at fixed prices from the farmers, usually leads to a short-circuiting of the local collecting trade functions of the small towns, while the farmers themselves are increasingly caught in a state monopoly that often excels in mismanagement and abuse. This negative impact of the rural development board system on local town hinterland relations has been widely observed in such different regions as Southern Mali (Van Teeffelen 1992, p. 194) and Central Java (De Jong & Steenbergen 1987; Titus 1991).

Finally, but certainly not least, both the supply and use of small town services are strongly determined by local contextual factors, such as geographical accessibility, population density and rural productivity or prosperity. In Southern Mali as well as in Bantul District (Java) considerable shifts in the supply and use of these services have occurred along with decreasing time-cost distances and changing levels of prosperity. Especially so, when these shifts are found in areas like Bantul District, where there is direct competition from a nearby higher order centre (Yogyakarta City), offering higher quality services. In that case it may even happen that the critical threshold value for maintaining basic public services in the small towns can no longer be attained, so that in the long run the poorer and less mobile sections of the rural population will be deprived from access to these services.

The Small Town Labour Market and Employment Role

Besides specific local differences, the structural characteristics of the small town labour markets in our case-studies also show some remarkable similarities. With the exception of the so-called *company* and *maquiladora* towns, the small town labour markets are all characterised by a large servicing sector which often seems to have absorbed more labour than strictly necessary for its proper functioning. Moreover, the small town labour market often is highly fragmented along both sectoral and occupational status lines, because of related differences in qualification requirements, wage levels, and occupational security. This leads to considerable differences in the accessibility of the respective labour market segments. The sizeable administrative and public services sector thus offers the most secure type of jobs, while the corporate enterprise sector in trade and industry usually offers the better paid jobs. But the great pressures exerted upon these segments of the urban labour market by local newcomers as well as the concomitant requirements of formal training, capital ownership and connections, make them relatively inaccessible to the majority of the small town and rural hinterland labour force. The most attractive of these urban labour market positions therefore, usually are occupied by higher-educated migrants from outside, whereas the local, less well-trained labour force has to content itself with the lower paid and insecure types of jobs in the family or individual enterprise sectors (cf. Romein in chapter 6).

Next to this vertical segmentation of jobs according to economic sectors, the small town labour market is stratified horizontally according to levels of income, education and employment status. Each occupational stratum or level of integration in the respective sectors of the labour market thus is characterised by specific entry requirements and differential accessibility for the respective labour force categories. Often these horizontal stratifications are more relevant than the sectoral divides. As Van der Wouden (1997) has shown for the Serayu Valley towns, the lowest level jobs for wage workers in the corporate sector often are hardly different in remuneration from those in the middle range of the family or individual enterprise sectors, which represent the petty-type activities. The relative scarcity of these lower corporate sector jobs and their somewhat higher level of income security and formal skill requirements however make them less accessible than the *open entry* petty-type jobs.

On the other hand, the large familial organised labour market sector (consisting of shops, small restaurants, cottage industries, domestic services, etc.) offers employment opportunities to independent entrepreneurs, dependent wage workers and unpaid family workers alike. Access to these opportunities usually is strongly determined by kinship and ethnic affiliations, especially in the Sub-Saharan African context where family enterprises operating according to modern capitalist principles are a relatively recent phenomenon (Van de Post 1988, p. 168). The small town entrepreneurs there often make use of unpaid family workers according to the *apprentice system* and have to accumulate their starting capital with the help of kin, who put them under heavy kinship obligations (cf. Hinderink & Sterkenburg, 1975). Consequently, few entrepreneurs are able to run their enterprises according to modern management rules, while their workers remain in a very dependent and exploited position.

The equally sizeable individual labour market sector of the small town offers employment opportunities to independent petty-type producers and casual workers in such different occupations like street seller, market trader, transport worker, repairer, carpenter or barber. Although there may be considerable differences in remuneration and job security among these individually pursued petty-type occupations, they distinguish themselves from occupations in the other labour market sectors by their relative *ease of entry* in terms of capital and formal training requirements. It is this ease of entry which causes the fierce competition among the petty producers in the individual sector through processes of overcrowding leading to low hourly wages, long working days and self-exploitation of labour. Ultimately, however, even within the individual labour market sector countervailing mechanisms may emerge which block further absorption of newcomers.

In our case-study towns these mechanisms were mostly related to the limited circulation of money and goods characterising the small town economy with its very modest corporate sector activities and its vulnerable hinterland relations. On the other hand, in the so-called *company towns* with their more prominent and dynamic corporate sector economy, petty-type activities often are discouraged by the local authorities due to their alleged non-modern and uncontrollable character (cf. Van der Post 1988, p. 183; Hinderink 1997, p. 184).

In summary, one might conclude that the small town labour market essentially shows the same structural characteristics as that of the large urban centres, but due to the absence of a dynamic corporate sector economy and weak hinterland relations it usually shows more signs of relative congestion and stagnation. The small town labour market therefore, also shows a high degree of horizontal and vertical compartmentalisation, as well as a pronounced pyramidal structure in which the higher and more secure positions are hardly available and accessible to the local labour force. Consequently, both labour mobility and absorption capacity in the small town labour market have to remain very limited.

A closer look at the absorption role of small towns with respect to their local and regional labour force, however, shows that there are considerable differences between the formally organised governmental and corporate sectors on the one hand, and the informally organised petty producer sectors on the other. In the first type relatively many migrants from outside the town's hinterland region are found with an advanced level of education, whereas the latter type mainly absorbs the lowly educated and unskilled labour force from the town itself and its immediate hinterland area. Labour mobility between the respective sectors remains limited and is mainly confined to the same integration level. Within the same type of sector however there often is some upward occupational mobility with rising levels of age and settlement duration, which suggests that the more dynamic towns offer opportunities for improving one's position, even within the petty producer sectors. The migration role of the small town, therefore, is mainly determined by the relative size and the dynamics of its corporate sector economy.

Not unexpectedly, the company towns of Swaziland and the *maquiladora* centres of Northern Mexico displayed the highest rates of net immigration, whereas the other types of centres with their limited local opportunities exerted much less attraction on migrants. Even within this latter category, however, there are still considerable differences as is demonstrated by the higher inmigration rates of Ciudad Quisada due to the advancing colonisation process in its hinterland area, versus the more or less stagnant centres of Mopti or Banjarnegara which only show low rates of natural increase.

Towns displaying a clear net migration growth also show different labour absorption patterns from stagnant centres. In the former, employment participation of migrants is relatively high in nearly all sectors of the labour market, whereas migrant participation in the latter remains limited to the government and corporate sectors. Here migration usually is connected with job transfers and official mutations, rather than with growing economic opportunities. For a proper understanding of the different integration opportunities for migrants in the various types of towns it is important to realise that a high degree of participation in the small town economy is not *per se* indicative of its favourable integration role, as it may concern very specific groups in limited sections of the town's labour market.

Finally, the small towns may also contribute indirectly to regional labour mobility through their educational functions. The concentration of secondary educational facilities in these centres usually also attracts pupils from their rural hinterland areas, who after school leaving often can not find suitable jobs or opportunities for advanced education and consequently are compelled to leave their home area for larger urban centres. In this way, the small towns in Sub-Sahara Africa and Central Java were obviously contributing to the drainage of human capital from their rural hinterland areas (Van der Post 1988, p. 158; Van Teeffelen 1993, p. 177; Van der Wouden 1997, p. 143).

The Regional Context of Small Towns

In the next sections an attempt will be made to elucidate the small town's role in developing its rural hinterland area under different conditions. This will be done by analyzing first the regional contexts of the various types of small towns, as these are assumed to play a major role in both their functioning and development potential (cf. chapters 1 and 2). Special attention will be paid to such factors as population density, accessibility, natural resource endowment and prevailing relations of production, next to the role of government policies with respect to rural and regional development. From our case-studies we could distinguish the following regional and policy contexts with respect to the small town's functioning:

- A Central Javanese or Southeast Asian context showing relatively densely populated rural areas with a favourable natural resource base characterised by intensive food and cashcrop farming systems and by peasant farmers who often supplement their meagre incomes with non-farm activities. Such areas usually produce relatively constant agricultural surpluses for the market, while on the other hand there is also a constant demand for agricultural inputs and urban consumer goods and services. Moreover, most of these areas are well served by infrastructural facilities like roads for motorized traffic, and by a relatively balanced system of larger and smaller towns offering a wide range of commercial and public services. Real big cities (>500,000) are rarely found in these areas or in their immediate neighbourhood, but certainly are making their influence felt on the rural economy as higher level centres of administration, marketing, distribution and employment for (circular) migrants of rural origin. Through increasing rural-urban interaction and the overall availability of non-farm employment opportunities and rural services, the rural economy has already experienced considerable diversification. On the other hand however development and prosperity may be severely hampered by widespread land fragmentation and landlessness, as well as by the marginal supply-push characteristics of most non-farm employment opportunities.

  Strong government intervention in rural and regional development issues are another prominent feature of the Southeast Asian context. These interferences usually are implemented through investments in infrastructural facilities (roads, irrigation, energy), Green Revolution programmes (food crops), educational facilities and health care. Besides, much attention is paid to institutional development in administration, marketing and production facilities (credit, extension, cooperatives). Small towns often are allocated a central role in the implementation of these policies at the local level, through the establishment of various types of public facilities and the reinforcement of their traditional servicing functioning in trade, commerce and processing (Van der Wouden 1997, p. 157; Huisman & Stoffers 1992). The impact of government policies on small town development however is not always a positive one in Southeast Asia. Price controls on food products or the abolition of subsidies on production inputs, often lead to deteriorating terms of trade for the agricultural producers. Moreover, government enforced obligations to cultivate certain cash crops (HYV rice or sugar cane) may reduce the rural purchasing power if this prevents the farmers from cultivating more profitable crops. In both cases rural-urban relations and the local small town functions are negatively affected, as is demonstrated in the case of the Serayu Valley towns (Van der Wouden 1997, p. 159).

- The Malian or Sahelian context showing less densely populated rural areas with mainly self-sufficient types of permanent foodcrop farming next to extensive types of cashcrop farming, animal husbandry and fishery on the basis of communally owned resources. These agricultural activities are mostly pursued on a very limited and ecologically unstable basis due to the prevailing semi-arid conditions. Infrastructural facilities in the rural areas are less developed than in the Southeast Asian context, although accessibility by truck and microbus is still at a reasonable level. The regional urban system moreover shows a rather balanced structure, consisting of old regional capitals, market towns, and local rural service centres with originally flourishing commercial and artisanal activities. In spite of these not too unfavourable conditions, both the rural and urban economies have been suffering from serious crises due to continued periods of drought and bureaucratic-economic mismanagement. The latter factor mainly manifests itself through the creation of unwieldy and costly state enterprises and marketing organisations, and through abortive attempts to increase agricultural production (Harts-Broekhuis & De Jong 1993, pp. 54-56). Because of the consecutive economic collapse, the region has become completely dependent on foreign food aid and scanty government budgets (mainly salaries) in order to maintain a minimum of local facilities. Consequently, rural-urban interaction is limited to very elementary types of services such as food distribution and offering shelter or (casual) work to the migrants/refugees of rural origin. The Southern Central Region of Mali, and especially the Mopti region provide the model cases for this context (cf. Van Teeffelen 1992, pp. 45 ff; Harts-Broekhuis & De Jong 1993, pp. 71 ff).

- The context of recently developed agricultural colonisation and export production areas in Central America (Costa Rica) and Southern Africa (Swaziland). These concern less densely populated rural areas, which have been developed since one or two generations by either independent settlers, or by large companies i.e. cattle ranches, plantations and mining companies. The first type of colonisation is mainly based on self-sufficient types of foodcrop farming, whereas the latter is based on extractive types of production for the world market. The natural resource position of the producers usually is relatively favourable due to the large average farm size and the land reserves still available (although frequently of marginal quality). On the other hand, accessibility and marketing facilities in these areas often are quite deficient, so that the marketing of agricultural surpluses and the distribution of production inputs and consumer goods are severely hampered, in particular for the small independent producers. The export-oriented extractive activities usually have a better and more direct access to markets, but these marketing channels mostly *bypass* the local market centres. As the urban system in recently occupied areas still shows vacancies, the regional servicing centres usually play a vital role for their hinterlands with respect to the marketing and distribution of local food products, production inputs, industrial consumer goods and public services. Often the position of these regional centres approaches that of semi-monopolistic urban centres, as there are hardly any competitors in a wide area. Consequently, and in spite of the forementioned limitations in accessibility, rural-urban interactions with these regional centres are quite intense. Moreover, advancing colonisation, increasing population pressure and rising surplus production in their rural hinterland areas, may further add to their functional dominance. Except for the construction of infrastructural facilities and the establishment of basic public services, government policy interferences usually play a minor role in recently developed colonisation areas. More obvious, however, are the impacts of world market price fluctuations and macro-economic policy measures on the local food and export crop production in these recently developed areas. The radical economic switch in Huetar Norte from cashcrop cultivation by pioneer farmers to extensive cattle breeding on large estates optimally demonstrates this point (cf. Romein & Schuurman 1990).

- The context of industrialising borderland areas in Northern Mexico, where recent processes of economic globalisation have completely transformed the regional economy. Until recently, this economy was dominated by extensive cattle breeding and grain farming next to intensive types of market gardening on irrigated plots. The small towns fulfilled regional servicing functions for a limited hinterland area that was producing increasing surpluses. This enabled these towns to develop a whole range of marketing and processing facilities which made them potentially attractive for other economic activities as well. At the same time a relatively balanced system of urban centres had evolved which laid the foundation for regional development processes of a more diversified type. The rising tide of economic globalisation and related interventions from the Mexican government soon brought an invasion of labour-intensive industries from the North in this borderland region. These industries were mainly established in the existing urban centres in so-called *maquiladora* or assembling plants and were attracted by cheap labour, infrastructural and tax exemption facilities, and last but not least by their strategic location vis-à-vis the United States market. The local resource base of the borderland region therefore has only contributed to its development in the initial stage of industrialisation, whereas presently the dynamics of the centres are mainly determined by external factors. Concomitantly, employment and income growth in these centres have become more or less detached from productivity levels in their rural hinterland. The latter, however, may profit from the increasing marketing opportunities for rural produce in the fast-growing *maquiladora* centres (Van Lindert & Verkoren, 1991).

Government policy obviously plays a decisive role in this type of industrialising regions by offering special facilities to foreign investors, improving the local infrastructure and supplying public services. These policies however mainly focus on the urban centres and therefore tend to neglect the rural hinterland areas.

The Small Town's Development Role

Not unexpectedly the regional and policy contexts outlined above strongly determine the small towns' development role with respect to their rural hinterland areas. By analyzing their role in terms of predominantly basic versus non-basic functions, and predominantly parasitic versus generative functions, it might be possible to arrive at more general conclusions about their regional development potential.

- In the densely populated Southeast Asian context, with its intensive types of agriculture and well-integrated rural market economy, the small town's functioning often appears to be of a supportive and non-basic type rather than of a generative or basic type. The main reason for this is that rural hinterland development through programmes for foodcrop and cash crop farming (Green Revolution), infrastructural facilities and public services is nearly always initiated and financed at the national or provincial level. Moreover, most executing institutions like marketing boards or credit banks are established either in the higher order centres or in the rural areas themselves, thus by-passing the regional service centres. The few basic-type activities established in these centres - mainly manufacturing industries - usually are highly dependent on external raw materials, technology and markets, and therefore hardly use local resources and services (except for labour). This implies that the small town servicing role is strongly determined by the size and productivity of its hinterland economy. At the same time the small town's servicing role is frequently threatened by a takeover from collecting and distributing traders in higher order centres, who are attracted by the increasing agricultural surpluses and rising purchasing power in the rural hinterland areas and who enjoy certain economies of scale over the small town traders. On the other hand, the emerging rural middle class is increasingly inclined to go to higher order centres for shopping and buying consumer durables, as these centres become more accessible with modern transport facilities and offer a wider choice to the consumers. Caught between the limitations of its rural hinterland potential and competition from higher order centres, the small town in the Southeast Asian context obviously can not afford an outright exploitative relationship with its hinterland area, lest it be by-passed. This leaves the conclusion that their contribution to regional development is mainly a supportive instead of a generative one. With respect to labour mobility the town's role might even be parasitic, as it tends to function as a transit station for locally educated school-leavers who can only find suitable jobs elsewhere.

- A very different picture emerges with respect to the small town's role in the Sahelian context. Here, the rural hinterland economy has collapsed under the combined onslaught of climatological and demographic processes of change and erratic policies. Consequently, the rural and regional service centres have been practically bereft of their hinterland functions, except for their basic role in food distribution and elementary public services. In this situation of economic contraction, the dominance of the local and regional service centres seems to have increased quite paradoxically, i.e. without a parallel reinforcement and diversification of their original servicing functions. Due to factors like the contracting flows of money and goods between these centres and their hinterlands, the deficient adjustment of their public services to the local needs and capacities of their hinterland populations, and the fact that as 'last resort centres' they tend to attract the more educated and venturing groups of these populations, the Sahelian small towns often display more parasitic than generative characteristics. This raises the interesting question whether this role may go into reverse as soon as climatological conditions improve again and enable rural surplus production. Considering the serious weakening of the small town production structure and the debilitating impacts of prolonged food aid and bureaucratic mismanagement, this is not a very probable option. The more so, since the role of the African small town is generally appraised as predominantly parasitic, even under 'normal' conditions of rural-urban interaction (cf. Schatzberg 1979; Kabwegyere 1979; Southall 1988; Pedersen 1991). The main reason for this is that the low productivity of the increasing urban population and its weak purchasing power, together with the ubiquitous concentration of political-economic power in urban centres, must lead to deteriorating terms of trade for the rural producers. In other words, the necessity to secure cheap supplies of food for the underproductive urban population is also negatively affecting the development of more balanced rural-urban relations (cf. Gooneratne & Obudho 1997, pp. 3-8).

- The role of the small town in recently developed colonisation and export-producing areas, tends to be much more positive. Not only because their production and servicing functions are more strongly developed, but particularly because these towns are fulfilling indispensable and even monopolistic functions for a wide and scarcely populated environment. Due to the absence of a balanced urban system and sufficient transport facilities, the rural hinterland producers are completely dependent on the nearest servicing centre for the marketing of their produce and for obtaining the necessary inputs and public services. In the case of the so-called *company towns*, which have been established with the prime purpose of serving the surrounding plantation and mining areas, there usually is a very strong dependency on a few extractive activities for the world market and related price fluctuations. By establishing processing industries for primary products some diversification of the small town economy may be achieved so that a larger share of the value added may be kept within the region. But ultimately, some net-surplus extraction remains inevitable due to the requirements of foreign 'holding companies' and national headquarters in the higher order centres or capitals. Moreover, purchasing power in the hinterland area also tends to be depressed by the narrow export base of the local economy, as well as by an unbalanced social structure consisting mainly of underpaid wage workers (cf. Gilbert & Gugler 1992, p. 50). A clear exception to this rule however is offered by export production areas which have developed on the basis of commercial types of agriculture by independent settlers. Here small towns usually perform an important role through the establishment of a range of administrative, public and commercial services supporting their booming hinterland area, as is illustrated in the cases of Swaziland and Huetar Norte (Costa Rica). In the case of Manzini the original company town features were gradually overruled by the latter type of hinterland development. Agricultural colonisation by independent settlers also played a decisive role in the development of Ciudad Quesada with its diversifying functional structure. The towns' more or less unique position as regional marketing and servicing centres together with their growing hinterland populations and increasing agricultural surpluses, pushed their economic structures and regional functions to a level beyond that expected on the basis of their sheer size. On the other hand speculative types of resource use may develop, which easily lead to processes of ecological degradation and 'hollow frontier' settlement. Under these circumstances, neither the required levels of population density, nor those for the intensification and commercialisation of agriculture will be achieved, so that there is little opportunity for the development of a balanced system of rural and urban service centres. Interesting cases of this type of frontier town development are found in the Amazon Region and in the Midwest of Brazil (cf. Volbeda 1997; Coy, Friedrich & Lücker 1997).

Finally, the position of the more recently established service centres may be eroded by changing economies of scale accompanying the introduction of export-monocropping systems in an originally mixed farming economy. This case is illustrated by Ciudad Quesada, which is increasingly suffering from bypass effects since the transformation of its mixed rural hinterland economy into a cattle export economy for the North-American beef market. The buying up of land by cattle ranches does not only lead to the eviction of the independent pioneer farmers and a switch to more extensive types of land-use, but also to a decline in the supply and processing of traditional cash crops (coffee, cereals), while the export of live cattle completely bypasses the town because processing and shipment take place in higher order centres without leaving any significant value added in its region of origin.

The structure and role of *'maquiladora'* towns, like Nueva Casas Grande in the Northern Mexican borderland area, is much less determined by their rural hinterland potential. The main reasons for this are that the rise of manufacturing activities and concomitant growth of services in these centres is nearly completely determined by foreign direct investments and by production relations focusing on the international market. Therefore, these centres may develop more basic functions and fulfil a stronger generative role towards their hinterland economies than the previous types of centres, but only if the hinterland economies have succeeded in obtaining advantage from the increased marketing opportunities for food and raw materials and from the new employment opportunities in these industrialising centres. For several reasons this often is hardly the case. Contrary to the positive impact of increasing local marketing and employment opportunities, there is the negative impact of increasing national and international competition through food imports and the influx of labour from other Mexican regions. Moreover, direct linkage effects between the rural hinterland economies and the industrial centres, usually are weak or absent as most manufacturing activities are based on the processing or assembling of imported raw materials and semi-manufactured articles. On the other hand, the relatively sizeable corporate enterprise sector in these centres, also generates a considerable circulation of money and goods which in its turn may give rise to local informal (petty-type) sector activities absorbing unskilled migrants from the hinterland areas and beyond (cf. Béneker, Van Lindert & Verkoren, 1997). This does not imply however, that the smaller *maquiladora* towns are really able to stem the flow of rural-urban migrants to the larger Mexican cities (cf. Béneker & Verkoren, in Chapter 7). For this purpose the quality of the employment offered usually is too low and one-sided as it is mainly limited to unskilled informal sector jobs and lowly paid female factory work. Moreover, demand for labour in these centres often fluctuates considerably, due to the predominance of 'footloose' industries which are very sensitive to local production conditions and international business cycles. With respect to the rural hinterland relations of these centres one may conclude however that their role on balance is a generative one due to the many basic types of industry and their positive effects on local employment and income levels which also stimulate the demand for rural hinterland products (mainly vegetables, fruits and daily products) and rural surplus labour. As far as functional relations among the *maquiladora* centres themselves are concerned, our case study could not offer any further insight. On the basis of scanty evidence from other studies one might expect these centres in the long run to develop stronger regional and inter-urban relationships through increasing labour mobility and exchange of commodities and services based on industrial supply and sub-contracting relations (cf. Van Lindert & Verkoren, 1991). But in the context of this globalising borderland economy, the development of a regionally integrated urban network might also be impaired by the strong international links of the local industrial activities. As most of them are completely dependent on foreign companies for capital, technology, inputs and markets, the opportunities for developing local and regional backward and forward linkages also have been seriously reduced.

The preceding discussion on the development role of different types of small towns, finally enables the construction of a scoring table on their most important hinterland functions. This table clearly demonstrates that centres with a relatively dynamic rural hinterland economy or with high levels of investment in their production structure show the best performance. It also shows that nearly all types of centres register low scores in processing local raw materials and offering regional employment opportunities.

Table 10.1: Scoring table on the strength of hinterland functions of small towns

Types of small towns	Types of hinterland functions					
	local processing	public services	marketing	distribution	employment	total + score
S.E. Asian service centres	±	+	±	+	-	3
Sahelian service centres	-	±	±	±	-	1.5
Newly developed area centres	±	+	±	+	±	3.5
Industrialising area centres	-	+	+	+	+	4

Some Theoretical and Policy Implications

In this section an attempt is made to draw some general conclusions from our case studies, which may be relevant to both theory and policy formulation on the role of small towns. This will be achieved by comparing the case-study findings with the results from more recent and general publications on this topic. The focus of our attention will be on the more or less universally accepted assumption among policymakers and regional planners that regional service centres play an essential role in rural hinterland development through direct production linkages and so-called 'spread' and 'trickling down' effects. Accordingly, the reinforcement of the small town production and institutional structures does not only contribute directly to rural and regional development, but is even seen as a necessary condition (cf. Rondinelli 1988, 1990; Evans 1992; Gaile 1992, p. 133).

Findings from our case studies and insights from other recent studies, however, often raise serious questions about this development role of small towns and often present ambiguities for the following reasons:

1. The inherent weakness of the small town production and servicing functions in the face of the capacities and needs of the rural hinterland population and its economy. Present facilities for processing raw materials or serving rural producers and consumers with inputs, consumer goods and marketing opportunities, often require 'economies of scale' and 'critical thresholds' that simply can not be provided or handled by small town establishments. This in turn leads to rather fragmented and inefficiently organised rural-urban trade relations, which then are increasingly 'by-passed' by more efficient traders operating from higher order centres. Considering the deteriorating terms of trade for agricultural produce vis-à-vis the commodities and services produced in these higher order centres, this would also imply an increasing transfer of surpluses produced in the rural hinterland areas to destinations outside the region.

   Moreover, the small town public services usually are mainly geared to the needs and capacities of the urban middle class, thus neglecting the needs of the rural and urban lower income masses and contributing to social and economic stagnation in their hinterland areas (cf. Van Teeffelen in Chapter 4; Huisman & Stoffers, 1992). These observations, which are particularly applicable to the first two types of towns mentioned in table 10.1, have been corroborated by other studies as well. In the African context the main causes are found in the small and insecure supplies of agricultural surplus (cf. Pedersen 1991, p. 246), as well as in the very low investments in infrastructural and public facilities at the lower end of the urban hierarchy (cf. Gaile 1992, p. 132). In the Southeast Asian context, with its much higher population densities, higher production surpluses and better developed transportation systems, the small towns frequently are being bypassed as centres of processing, marketing and distribution, due to the competitive advantages of higher order centres. Mainly for that reason, according to Evans (1992, p. 665), large corporate enterprises tend to avoid the smaller centres, while the locally overrepresented petty-type enterprises usually generate only few forward and backward linkages.

As far as the trade relations of the small towns are concerned, Pedersen (1991, p. 247) also points at the strong fragmentation of the trade activities, as well as at the prevailing patron-client relationships. Although such relationships serve to guarantee the necessary flexibility and stability under conditions of insecure supplies, it may also lead to undesirable forms of dependency and surplus extraction among the producers, while the great number of trade links lead to high transaction costs. Consequently, the small town trade relations are under constant pressure from both the more efficiently operating large-scale traders in higher order centres and the rural hinterland producers seeking better prices for their surpluses.

2. The fact that rural hinterland development of small towns derives its impetus and momentum from macro-policies and macro-economic processes at the national level, rather than from local impulses emanating from the towns themselves. This concerns especially the impact of rural development programmes focusing on the Green Revolution, small-scale industries, infrastructural facilities and public services, all of which are directly and indirectly affecting the rural production levels and living standards and consequently, the demand for urban goods and services. It should be kept in mind, however, that government policies do not always affect rural hinterland development in a positive way. Such may be the case for example in the introduction of monocropping systems for export production, or the imposition of cash-crop cultivation for national self-sufficiency. Both the introduction of commercial groundnut and cotton cultivation in Sub-Sahara Africa (cf. Jamal 1995, p. 20) and the compulsory cultivation of HYV rice and sugar cane in Indonesia (Titus et al. 1994) have had depressing effects on rural income levels, as farmers were prevented from cultivating more secure or more rewarding types of crops. In such instances the development of the local rural service centres may be negatively affected as well, as has been demonstrated in the Serayu Valley and Bantul District cases.

The incorporation of rural hinterland areas in the national and international market economies may also proceed more or less spontaneously and without the mediation of local service centres. The resulting growth and diversification of the hinterland economy then usually stems from increasing marketing and employment opportunities in other regions and higher order centres. The latter process usually is triggered by improved rural-urban communications, facilitating permanent and circular types of labour mobility to especially the larger cities. These processes have been widely observed and documented in the densely populated and commercialised rural areas of South and Southeast Asia where they may contribute to more than 50% of the rural household incomes (cf. Hugo 1996; White 1991; Bhalla 1995; Harriss & Harriss 1987; Trager 1988).

In Latin America these externally induced forms of rural hinterland incorporation and development often are related to agricultural colonisation areas which focus on the production for national and international markets, such as Amazonia (Volbeda 1997), Huetar Norte (Romein 1995) or the Santa Cruz area in Bolivia (Zoomers 1997). The case of Ciudad Quesada however has shown that this type of development does not always imply the reinforcement of the local servicing town's position.

In Sub-Saharan Africa similar cases of externally induced rural development have been documented for areas like Western Kenya, Zimbabwe and Swaziland (cf. Evans 1992; Pedersen 1997). But in general rural-urban and farm-non farm relations there are much less developed than in Asia or Latin America, due to higher degrees of rural self-sufficiency and lower levels of productivity (cf. Haggblade et al 1989). Moreover, small town functions often are ill-adapted to the needs of the rural hinterland population, so that small town development is only remotely connected with hinterland development.

From the above synopsis one may conclude that small towns rarely play a prominent role in starting rural development in their hinterland areas. At most they seem to fulfil a supportive role, but in the light of the predominant national and international factors, this role is only of minor importance.

3. The fact that changes in the structural characteristics and the role of small towns tend to follow the rural hinterland developments described above rather than initiate them. It is generally accepted now that rural service centre development is only possible with growing rural surplus production and marketing. Mellor (1990, 1995), for example, has pointed at the vital role of agricultural intensification and commercialisation in creating this rural surplus, which in turn generates a strong demand for urban goods and services. This holds not only for consumption, but also for productive purposes like credit, inputs, transport and repair. This demand arises in particular from the emerging rural middle class, consisting of richer farmers and traders who spend a larger part of their income on so-called non-food commodities and services.

Other authors, however, have pointed at the fact that increasing rural surplus production and demand do not always engender higher expenditures in the local service centres or towns (cf. Dunham 1991, pp. 6-9). Especially the higher income groups tend to buy consumer durables and imported luxury items which do not support the local hinterland economy and which usually are obtained in higher order centres outside the region. Inversely, they may directly sell their larger surpluses to nearby higher order market centres, thus short-circuiting the local market centres and obtaining better prices.

Another point of doubt concerns the assumed increase of productive investments in non-farm activities with rising rural incomes, which in turn may prompt the diversification of both the rural and small town economies (cf. Hinderink & Titus 1988). In the case of deficient local production facilities and under conditions of great climatological and economic insecurity or speculative types of resource use - such as in the African Sahel zone or in the Latin American frontier areas - the expected investment responses may simply fail to materialise and the growth of small towns will lag behind. The small town production structure moreover may also be constrained by direct government interventions at the national level, such as by structural adjustment programmes based on considerable budget cuts and deregulation measures. Retrenchments in government expenditures tend to come down hard on the small town economy, due to the town's large and relatively dominant public sector functions which are struck by subsequent wage freezes, redundancies and the postponent or abandonment of public schemes (cf.

Holm 1995, p. 100; Bryceson 1996). Simultaneous deregulation measures may also have negative short-term impacts for the few corporate enterprises in town due to the concomitant rises in interest rates, the lifting of market protection and increased competition from imported goods. These impacts of structural adjustment policies have been corroborated by findings from our case-studies in Huetar Norte (cf. Romein 1995) and Central Java (Van der Wouden 1997), where the small town economies experienced severe dips in the mid-1980s. Pedersen (1997, pp. 164-165) however also observes contrary developments in Zimbabwe, where the long-term impacts of structural adjustment policies on small and medium size enterprises seem to have been mainly beneficial. Improved access to credits, raw materials and spare parts for the smaller enterprises and a higher degree of flexibility towards market shifts have given these enterprises a definite advantage over the formerly protected corporate enterprises.

In summary we may conclude therefore that in every respect the national and regional economic and policy contexts have been more important determinants of the small town's structural characteristics and dynamics than might be assumed on the basis of the town's 'pacemaker' role towards its hinterland.

4. There is strong evidence that next to the impact of the previously discussed national and macro-structural policies and developments the structure and role of small towns are mainly determined by local and regional economic conditions. Among these local economic conditions we may distinguish such obvious factors like natural resource endowment and population density, market accessibility, types of production relations and distribution of income and purchasing power. The role of the natural resource factor is particularly evident in areas with uncertain rainfall conditions and precarious agricultural yields like the Sahel. The insecurity which this inflicts upon the farmers' lives has a direct impact on their spending and investment behaviour and consequently also on their demand for urban goods and services. This is exemplified by Mopti's economic stagnation and its inability to develop a corporate sector economy based on the processing and marketing of rural hinterland products (cf. Harts-Broekhuis & De Jong 1993, p. 93). A similar phenomenon may occur in agricultural colonisation areas with unstable soil conditions, such as in tropical rainforest areas with lateritic soils. Here quick soil depletion usually stimulates a shift to more speculative and extensive types of land-use and subsequent environmental degradation. Moreover, the transformation of subsistence types of pioneer farming into extensive types of commercial farming or cattle ranching may easily lead to 'hollow frontier' conditions and stagnant service centres (cf. Coy, Friedrich & Lücker 1997, pp. 31-52).

Sometimes large-scale exploitation of local minerals and other raw materials (timber) is the most decisive factor in small town development. The so-called 'company towns' often have been founded with capital and support from mining and plantation companies, and therefore accommodate modern facilities of a much higher quality than is usual for service centres of their rank and size. These facilities however may have little or no function at all for the majority of the rural hinterland population because they are mainly serving the needs of the companies and the better paid employees and civil servants, while neglecting the wage workers and poor subsistence farmers. As soon as the local resources have been depleted or substituted by the world market, the towns concerned will fall into decay and their position is reduced to the level of stagnant rural service centres (Volbeda 1984, pp. 151-153; Evans 1989; Hinderink 1997).

Still more important is the role of the human resource factor expressed by variables like population density, growth and productivity. Under conditions of low population density and low productivity like in the Latin American frontier areas or in Sub-Saharan rural Africa, it may become extremely difficult to develop an adequate physical infrastructure and generate sufficient local surplus production. Consequently, the distribution of service centres may be very spread out, resulting in large areas being left unserviced and a poorly developed urban hierarchy. Because of lack of competition among the centres and their low levels of specialization, most of these centres offer a wide variety of servicing functions but usually only of low quality (Van Teeffelen 1992, pp. 199-200; Volbeda 1997). Conversely, higher rural population densities and levels of productivity like in the Southeast Asian context, may sooner lead to the formation of dense networks of service centres, due to a larger local demand for urban commodities and services, as well as a higher supply of locally produced surpluses for the market. Even in cases where only part of the rural population generates sufficient demand and surplus production, the critical threshold values for sustaining small town development will be attained earlier in the densely populated areas. This is clearly demonstrated in the Central Javanese context where in spite of rural per capita production and income levels being among the lowest in the country, this has not prevented the formation of a dense network of rural and regional service centres (cf. Huisman & Stoffers 1992; Van der Wouden 1997, pp. 57-58).

Another important factor contributing to the positive relationship between rural population pressure and small town development may arise from the necessity for the rural surplus labour to create (additional) employment in the non-farm sector. These small-scale, so-called 'supply-push' types of non-farm activities usually show a preference for establishment in rural service centres due to their locational advantages with respect to nearby markets, transport and energy facilities, and raw material supplies (cf. Dunham 1991, p. 23; Berry 1995, p. 294). Not unexpectedly, this positive relationship appears to be strongest in the Southeast Asian and Latin American contexts where rural population densities, income levels and demand for non-agricultural products reach higher levels and farm-non farm linkages are more direct than in Sub-Saharan Africa (cf. Haggblade et al 1989, p. 29; Pedersen 1997, p. 12).

Obviously, local and regional differences in market, production and consumption conditions play a vital role in shaping the small town's economic structure and functions. Market accessibility, for example, does not only determine the level of commercialisation of the hinterland economy, but also the range of the towns' servicing areas and their competitive relations. In cases of overlap between the servicing areas, so-called 'bypassing' and 'erosion' effects may occur which ultimately lead to a loss of functions in the weaker centres. This impact was particularly obvious in the case of the Central Javanese towns, which do not only experience heavy competition from higher order centres in their collecting and distributive trade functions, but also from the small towns among them like in the Serayu Valley region (Titus 1991; Van der Wouden 1997, pp. 158-159).

As the small town economy is not only determined by factors of market accessibility and rural hinterland productivity, but also by various administrative structures and institutional arrangements, the role of political-economic factors becomes particularly relevant for explaining the differences in development opportunities. Particularly the control of investment flows and public finances may offer relevant explanations for the frequent observation that regional administrative centres (e.g. district capitals) tend to expand or consolidate their economic functions at the expense of the lower order centres in their hinterland areas (cf. Gaile 1992). Government investments and bureaucratic control indeed seem to have played a major role in the development of regional centres like the Serayu Valley towns, the Swaziland towns and the Malian towns (i.c. Mopti). Local relations of production are another important political-economical aspect determining the rural hinterland productivity and rural-urban relationships. In rural areas with independent farmers producing food and industrial cash crops, Green Revolution programmes for example, will have a much more favourable impact on local productivity and local demand for urban commodities and services, than in areas characterized by large landownership or *latifundia* with (share) tenants. In the first case surplus production mainly accrues to the farmer-tillers, whereas in the latter case it mainly accrues to the landowners and the traders supplying credit to the impoverished tenants in exchange for low off-farm prices. Experiences with Green Revolution programmes in divergent areas like the Punjab (India), Java (Indonesia), Central Luzon and the Visayas (Philippines) indeed offer strong evidence in support of this differential impact on rural development. In the Philippines, and especially in the *hacienda* areas, the Green Revolution was less successful and income distribution more skewed than in Java or the Punjab, so that rural and regional service centre development there could take much less advantage of an increasing rural demand for consumer goods, inputs and services (cf. Bautista 1995; Boyce 1993).

Similar impacts have been reported from different areas in Java, whose western province has a less egalitarian land tenure system than the other two provinces. Here absentee ownership and sharecropping arrangements have brought lower yields to the farmer and resulted in weaker linkage effects of rising rice production with non-farm production (Sadoko 1989, pp. 46-47).

Moreover, since areas more or less exclusively devoted towards large scale export crop production by a labour force of unskilled wage workers, usually offer fewer opportunities for developing a balanced urban network (cf. Gilbert & Gugler 1992, p. 50), rural service centres frequently are underdeveloped. This is not only due to local conditions of low purchasing power and lacking demand for production inputs and marketing facilities by independent farmers, but also to the fact that large scale agricultural or cattle production for the export market needs only support from a few larger centres or company towns. This situation applies in particular to the vast corporate type plantation areas in Southeast Asia and Central America, as well as to the extensive cattle ranching areas in South America, but it might equally apply to former 'white settler' areas in Sub-Sahara Africa, like in Kenya, Swaziland or Zimbabwe (cf. Van der Post 1988, pp. 84-85; Wanmali 1996).

In the more densely populated tribal or communal farming areas in Sub-Sahara Africa however, the prevalence of self-sufficient and stagnant types of foodcrop production offers little impetus for rural service centre development. Usually there is a network of local service centres but only in a rudimentary form providing basic services and commodities. As most service centres are not able to perform productive functions for their stagnant hinterland areas by processing and marketing rural surpluses, the remaining distributive functions often are of a parasitic type (cf. Southall 1988; Pedersen 1990; Evans 1992). By selling mainly imported commodities and food at deteriorating terms of trade to their hinterland populations, these centres indeed are draining the small rural surpluses. Moreover, by providing urban-oriented educational facilities to their hinterland populations, these centres often are also contributing to a continuous 'brain drain' from these areas. The main reason for the negative role of African small towns in rural development, therefore, might be their inability to bridge the gap between a predominantly self-sufficient and communally organized rural economy on the one hand, and a market-oriented capitalistic economy at the national level on the other. As soon as rural relations of production become more commercialised and rural purchasing power rises, small towns may also perform a stimulating role in the African context, as has been exemplified by the case-studies from Swaziland (cf. Van der Post 1988. pp. 272-275). This view takes us to our final policy relevant conclusion.

5. With respect to the role of small towns, the dialectical course of rural-urban interaction in regional development (see chapter 1) remains confined to areas where rural relations of production already have been transformed to the extent that the majority of the producers are connected with the national and international markets. This does not exclude however the fact that the rural economy has been integrated in a subordinated and exploited position, finding expression in unfavourable terms of trade for foodcrops, or in a growing dependency on export-oriented cropping systems which are subject to strong price fluctuations. It is essential though, that there should be some net increase of the rural surplus production, e.g. through rising yields per unit area which compensate for possible price losses. Moreover, at least part of the produced surpluses should remain in these rural areas and their service centres in order to enable private investments into local economic growth and diversification (cf. Hinderink & Titus 1988; Pedersen 1997). This does not only require a more or less egalitarian and owner-based rural production structure as signalled by Southall (1988, p. 5) and Harriss & Harriss (1988) for respectively the African and Indian contexts, but also a relatively strong and indispensable position of the rural producers in securing national food supplies and earning foreign currency.

In addition, convincing evidence has been found that the small town functions should be backed by an emerging modern middle class of entrepreneurs, professionals and civil servants, who have the training, knowledge and the means to exploit new opportunities and develop more efficient hinterland relations (cf. Harriss & Harriss 1988). This is corroborated by both negative and positive examples in our case-studies. In cases of a lacking or poorly developed urban middle class, like in Mopti or Banjarnegara, the urban functions often were poorly performed and of low quality, whereas the presence of ethnic minority merchants and entrepreneurs in many Southeast Asian and African towns clearly contributed to their dynamic functioning (cf. Van der Wouden 1997, pp. 152-153; Van der Post 1988, p. 237).

Only if these conditions are fulfilled, some balance will be achieved in the basically unequal rural-urban relationships, as rural producers now may enjoy a certain bargaining position with the dominant forces of the market and the government, while the small towns may play a mediating role in bridging the conflicting interests between the local, regional and national levels. At the same time, small towns may enjoy some intrinsic advantages over larger urban centres due to the functional intertwining with their own hinterland interests. Consequently, they can not afford extremely exploitative hinterland relations, if they do not want to be by-passed by the more efficiently producing higher order centres. Besides, for reasons of proximity, rural and regional service centres often seem to enjoy initial advantages over higher order centres as locations for non-farm activities arising from local initiatives and (re-)invested surpluses (Pedersen 1997, p. 2).[1] The foregoing implies that governments primarily should focus their sectoral and regional development efforts on generating agricultural and non-farm surpluses by improving the rural physical and institutional infrastructures and marketing facilities, as well as by supplying credit, inputs, extension services and price incentives to the farmers. Part of this rural development effort may be channelled through the regional and rural service centres, but much can be achieved directly at the village and farmer levels by providing local services and incentives.

From the preceding discussion one may conclude that it becomes only justifiable to divert public means to the reinforcement of the small town structure and functioning (e.g. by investing in market centres, energy and water supply, or in housing and industrial estates) if certain basic rural development conditions have been met. Thus, there seems little in support of a possible reversal in the regional development efforts as suggested by Rondinelli (1983, 1988), since hardly any evidence has been found for a really initiating role of small towns in rural hinterland development. The more obvious however is the small town's supportive role in rural hinterland development as soon as the required conditions for this development have been fulfilled at the local and higher levels of decision making and policy implementation. It is in creating these conditions and in stimulating this supportive role that the main justifications are found for a planned effort aimed at rural hinterland and rural service centre development.

A crucial point in question is whether these rural and small town development efforts should be primarily defined in terms of regional or sectoral policies. In practice a sectoral approach seems much easier to implement as it requires less complicated coordination and achieves more predictable and short-term measurable impacts in other sectors through direct linkage effects. Programmes for agricultural intensification and rural economic diversification, for example, usually have a direct positive impact on rural service centre development by increasing the tradable surpluses of both farm and non-farm activities and by raising the demand for urban commodities and services. The successfulness of sectoral programmes however is strongly determined by local resource conditions and production relations, and consequently also their impact on the small town economy. This means that successful sectoral planning requires the consideration of regional conditions and should be carefully adapted to these local conditions. With respect to small town development this implies, for example, the avoidance of mechanical planning models allocating public servicing facilities according to rules based on central place theory (cf. Wanmali & Islam 1995). As our case studies have demonstrated both the nature and quality of these facilities should be better adapted to the needs and capacities of both the local urban and rural hinterland populations (cf. Van Teeffelen 1992; Huisman & Stoffers 1992).

In cases where town-hinterland relations are included in a regional development planning approach, the regional political-economic context becomes even more relevant. Thus, completely different approaches are required for reinforcing small town functions in the Southeast Asian context or in the Sahelian context. In the first context it is feasible to strengthen the small town functions which serve the growing demand from a densely populated and increasingly productive rural hinterland area that is not exclusively dependent on its own service centre. This means that the small town functions should be reinforced in such a way that more surplus is kept within the region, for example, by improving its distributive and collecting trade functions, establishing local processing industries, etc.

In the Sahelian context there are few reasons for stimulating the demand side of the regional economy by reinforcing the small town's commercial functions. Instead, all efforts should be directed towards improving the supply side of the stagnant rural hinterland economy and improving the town's poorly developed public service functions *vis-à-vis* the needs of its impoverished hinterland population.

Similar differences in regional conditions should be observed when dealing with town-hinterland relations in pioneer farming areas or in export-oriented *hacienda* areas. In scarcely populated pioneer farming areas problems of accessibility, marketing and public services usually are most acute, which implies that most measures should be aimed at reinforcing small town functions which serve the demand side of the regional economy, whereas in the *hacienda* or *latifundia* areas there still is much scope for improving the supply side by intensifying and diversifying land-use and by changing obsolete production relations preventing an economical use of available resources and urban functions. It goes beyond the scope of this book, however, to present a complete discussion of the various policy implications of the different regional settings of town-hinterland relations. This would require much more empirical evidence than presently is available, if such an elusive topic as rural-urban relations may ever become a practical tool for regional development planning intervention.

Note

1. With respect to the African situation Pedersen (1997) however notices that the establishment of non-farm activities and services in rural service centres is not an automatic response, since their location is strongly determined by the local impacts of sectoral development programmes, the supply-push character of most of these activities and the competition from higher order centres.

Bibliography

Abdullah, F. & T. Etty
1995 'Would Be' and 'Make Believe' in Crisis. In: D.R. Harriss (ed.) *Prisoners of Progress. A Review of the Current Indonesian Labour Situation*. Meppel: INDOC Krips Repro.

Adamchak, D.J.
1978 Migration to a Small Urban Place: an Examination of Migration Histories to Creel; Chihuahua; Mexico. *Estadística* 32, pp. 11-17.

Aguilar, I. & M. Solis
1988 *La elite ganadera en Costa Rica*. San José: UCR.

Ahmed, A.G. & M.A. Rahman
1979 Small Urban Centres: Vanguards of Exploitation. Two Cases from Sudan. *Africa* vol. 49 (3); pp. 258-271.

Altenburg, T., W. Hein & J. Weller
1990 El desáfio económico de Costa Rica: desarrollo agro-industrial autocentrado como alternativo. San José: DEI.

Amin, Samir (ed.)
1974 *Modern Migrations in Western Africa*. International African Institute, Oxford: Oxford University Press.

Anderson, D. & W.H. Leiserson
1980 Rural Non-Farm Employment in Developing Countries. *Economic Development and Cultural Change* 28 (2), pp. 227-248.

Andriessen, B. & K. van den Broek
1988 *Small Town Enterprises and Regional Development: The Case of Purworejo-Klampok, Central Java*. Diskussiestukken van de Vakgroep Sociale Geografie van de Ontwikkelingslanden, nr. 38, University of Utrecht.

Appalrayu, Jaya & Michael Safier
1976 Growth-Centre Strategies in Less-Developed Countries. In: A. Gilbert (ed.). *Development Planning and Spatial Structure*. London: Hutchinson; pp. 143-167.

Armstrong, W. & T.G. McGee
1985 *Theatres of Accumulation: Studies in Asian and Latin Urbanisation*. London: Methuen.

Bell, M.
1986 *Contemporary Africa: Development, Culture and the State*. Essex: Longman.

Béneker, T.
1997 *'Buscar Mejor Ambiente' Migratie naar, uit en langs een kleine stad in Costa Rica*. Dissertation, Utrecht: KNAG/NGS.

Béneker, T. & A. Hooijmaijers
1990 *Het functioneren van migranten op de arbeidsmarkt van Ciudad Quesada, een kleine stad in Costa Rica*. Diskussiestukken van de Vakgroep SGO 50. Utrecht: Faculteit Ruimtelijke Wetenschappen.

Béneker, T., P. van Lindert & O. Verkoren
1997 Migrant-Native Towns. In: Lindert, P. van & O. Verkoren (eds.), pp. 101-110.

Béneker, T. & A. Romein
1994 Migration and Migrant-Labour Absorption in Lower Order Urban Centres in Latin America: Two Regional Cases from Costa Rica. In: A. Harts-Broekhuis & O. Verkoren; pp. 331-341.

Berlinck, M.T., J.M. Bovo & L.C. Cintra
1981 The Urban Informal Sector and Industrial Development in a Small City: the Case of Campinas, Brasil. In: Sethuraman, S.V., *The Urban Informal Sector in Developing Countries: Employment, Poverty and Environment*, Geneva: ILO, pp. 159-167.

Berry, A.
1995 The Contribution of Agriculture to Growth; Colombia. In: J.W. Mellor (ed.), *Agriculture on the Road to Industrialization*. Baltimore/London: Johns Hopkins University Press, pp. 276-294.

Berry, B.J.
1972 Hierarchical Diffusion. The Basis of Developmental Filtering and Spread in a System of Growth Centres. In: P.W. English & R.C. Mayfield (eds.). *Man, Space and Environment.* Oxford: Oxford University Press; pp. 340-359.

Bhalla, G.S.
1995 Agricultural Growth and Industrial Development in Punjab. In: J.W. Mellor (ed.), *Agriculture on the Road to Industrialization.* Baltimore/London: Johns Hopkins University Press, pp. 97-107.

Bijlmer, J.
1987 *Ambulante straatberoepen in Surabaya. Een studie naar kleinschalige aktiviteiten.* Amsterdam: Vrije Universiteit, Dissertation.

Bilsborrow, R.E. & R. Fuller
1988 La selectividad de los emigrantes rurales en la Sierra Ecuatoriana. *Estudios Demográficos y Urbanos* 3-2, pp. 265-290.

Bilsborrow, R.E., T.M. McDevitt, S. Kossoudji & R. Fuller
1987 The Impact of Origin Community Characteristics on Rural-Urban Out-Migration in a Developing Country. *Demography* 24-2; pp. 191-210.

Bilsborrow, R.E.; A.S. Oberai & G. Standing
1984 *Migration Surveys in Low-Income Countries: Guidelines for Survey and Questionnaire Design.* London: Croom Helm.

Binsbergen, W.M.J. & H.A. Meilink (eds.)
1978 *Migration and the Transformation of Modern African Society. African Perspectives*, vol. 1. Leiden: Afrika Studiecntrum, vol. 1.

Birkbeck, C.
1979 Garbage, Industry, and the 'Vultures' of Cali, Colombia. In: R. Bromley, R. & Ch. Gerry (eds.), *Casual Workers and Poverty in Third World Cities*, Chichester: John Wiley & Sons, pp. 161-183.

Blau, D.M.
1986 Self-Employment; Earnings and Mobility in Peninsular Malaysia. *World Development* vol. 14 no. 7; pp. 839-852.

Booth, A.
1986 Survey of Recent Developments. *Bulletin of Indonesian Economic Studies* Vol. 22 no. 3; pp. 1-26.

Booth, A. & K. Damanik
1989 Central Java and Yogyakarta: Malthus Overcome? In: H. Hill (ed). *Unity and Diversity. Regional Economic Development in Indonesia Since 1970.* Singapore: Oxford University Press.

Booth, A.R.
1982 The Development of the Swazi Labour Market, 1900-1968. *South African Labour Bulletin* vol. 7 no. 6, pp. 34-57.

Borsdorf A.
1981 Probleme und Chance Lateinamerikanischer Mittelstädte. In: *Zweiter Tübinger Gespräch zur Entwicklungsfragen,* Universität Tübingen: Geografisches Institut; pp. 143-153.

Boyce, J.
1976 *Een dualistisch arbeidsbestel?: een kritische beschouwing van het begrip de informele sector.* Rotterdam: Van Gennep.
1980 The Informal Sector in Research: Theory and Practice. In: *CASP III.* Bromley, R. (ed.).
1985 *Planning for Small Enterprises in Third World Cities.* Oxford: Pergamon Press.
1993 *The Philippines: The Political Economy of Growth and Impoverishment in the Marcos Era.* Houndsmills/Basingstoke: McMillan.

Bradshaw, S.
1995 Female Headed Households in Honduras. *Third World Planning Review* 17; pp. 117-132.Breman, J.C.

Breman, J.C.
1976 *Een dualistisch arbeidsbestel?: een kritische beschouwing van het begrip de informele sector.* Rotterdam: Van Gennep.
1977 Labour Relations in the 'Formal' and 'Informal' Sectors: Report of a Case Study in South Gujarat, India,

        Journal of Peasant Studies, vol. 4, no. 3, pp. 171-205 and pp. 337-359.
1980   The Informal Sector in Research: Theory and Practice. In: *CASP III*. Bromley, R. (ed.).
1985   A Dualistic Labour System? A Critique of the 'Informal Sector' Concept. In: Bromley, R. (ed.), *Planning for Small Enterprises in Third World Cities*, Oxford: Pergamon Press, pp. 43-64.
Bromley, R.
1978   The Urban Informal Sector: Why is it Worth Discussing? *World Development*, vol. 6, no. 9/10, pp. 1033-1039.
1984   The Urban Road to Rural Development: Reflections on USAID's 'Urban Functions' Approach. In: Detlef Kammeier, H. & P.J. Swan, *Equity With Growth*, Bangkok: A.I.T., pp. 378-383.
1988   Working in the Streets: Survival Strategy, Necessity or Unavoidable Evil? In: J. Gugler (ed.) *The Urbanization of the Third World*. Oxford: Oxford University Press.
Bromley, R. & Ch. Gerry
1979   Who Are the Casual Poor? In: Bromley, R. & Ch. Gerry (eds.), *Casual Work and Poverty in Third World Cities*. Salisbury: John Wiley & Sons, pp. 3-23.
Bromley, R. & C. Gerry (eds.)
1979   *Casual Work and Poverty in Third World Cities*. New York: John Wiley & Sons.
Bryceson, D.F.
1996   De-agrarianization and Rural Employment in Sub-Saharan Africa; A Sectoral Perspective, *World Development* 24, pp. 97-111.
Cantú Guttiérrez, J.J. & R.L. González
1990   Migración a la zona metropolitana de la ciudad México. *Demos* 17-18.
Castles, S. & M.J. Miller
1993   *The Age of Migration. International Population Movements in the Modern World*. London: MacMillan.
Castells, M. & A. Portes
1989   World Underneath; the Origins, Dynamics, and Effects of Informal Economy. In: Portes, A., M. Castells & L.A. Benton (eds.), *The Informal Economy, Studies in Advanced and Less Developed Countries*, Baltimore/London: Johns Hopkins University Press, pp. 11-37.
Censo General de Población y Vivienda 1970
1971   Mexico DF: DGE.
Censo General de Población y Vivienda 1980
1983   Mexico DF: INEGI.
Censo General de Población y Vivienda 1990
1991   Mexico DF: INEGI.
Census
1986   *Provisional Data of the 1986 Swaziland Census* (unpublished).
Central Statistical Office
1987   *Population & Housing Census Analytical Report* Gaborone.
Chandler, G.
1984   *Market Trade in Rural Java*. Monash University Centre of Southeast Asian Studies. Melbourne.
Chandra, R.
1992   *Industrialisation and Development in the Third World*, London/New York: Routledge, 124 pp.
Cissé, S. & P.A. Gosseye
1990   Compétition pour des Ressources limitées: le cas de la cinquième région du Mali, rapport 1: ressources naturelles et population. Mopti/Wageningen: ESPR/ Centre for Agrobiological Research.
Collier, W.L. et al.
1982   Acceleration of Rural Development in Java. *Bulletin of Indonesian Economic Studies* 18 (3); pp. 84-101.
Coquerie-Vidrovitch, C.
1988   *Africa, Endurance and Change South of the Sahara*. Berkeley: University of California Press.
Coy, M., M. Friedrich & R. Lücker
1997   Town and Countryside in the Brazilian Midwest. Modernization and Urbanization of a Pioneer Region. In: Lindert, P. van & O. Verkoren (eds.), pp. 31-52.

Crush, J.
1982 The Southern African Regional Formation: A Geographical Perspective. *Tijdschrift voor Economische en Sociale Geografie*, vol. 73 (4); pp. 200-212.

Davies, R.
1979 Informal Sector or Subordinate Mode of Production? A model. In: Bromley, R. & Ch. Gerry (eds.), *Casual Work and Poverty in Third World Cities*, Salisbury: John Wiley & Sons, pp. 87-104.

Davila. J. & D. Satterhwaite (eds.)
1987 *The Role of Informal Sector in the Development of Small Towns in Rural Regions*, Paper presented at the conference on the informal sector as an integral part of the national economy, research needs and aid requirement, Copenhagen, September 1987.

Dewar, D., A. Todes & V. Watson
1986 *Regional Development and Settlement Policy. Premises and Prospects.* London: Allen & Unwin.

Dewey, A.G.
1962 *Peasant Marketing in Java.* New York: Free Press of Glencoe.

DGEC (Dirección General de Estadística y Censos)
1965 Censo de Población 1963, San José: Ministerio de Economia, Indústria y Comercio.
1975 Censo de Población 1973, San José: Ministerio de Economia, Indústria y Comercio.
1987 Censo de Población 1984, San José: Ministerio de Economia, Indústria y Comercio.

DHV
1985 Aspects of Rural Centre Planning, vol 1, *Theoretical Considerations*, Amersfoort.

Diagnostic Régional
1985 *Edition Définitive*. Bamako/Mopti: Comité Régional de Développement.

Dick, H.W. & D.J. Rimmer
1980 Beyond the Formal/Informal Sector Dichotomy Towards an Integrated Alternative. *Pacific Viewpoint*, vol. 21, no. 1, pp. 26-41.

Dijk, M.P. van
1980 *De informele sector van Ouagadougou en Dakar. Ontwikkelingsmogelijkheden van kleine bedrijven in twee Westafrikaanse hoofdsteden.* Amsterdam: Free University, 340 pp.
1982 The Role of Small-Scale Enterprises in the Development Process: A One Week Seminar on the Informal Sector. *Collaboration Series* no. 1; Universitas Satya Wacana; Salatiga;Amsterdam: Vrije Universiteit.

Diop, M.
1971 *Histoire des classes sociales dans l'Afrique de l'Ouest, le Mali.* Paris: Maspéro.

Djajadiningrat-Nieuwenhuis, M.
1987 Ibuism and Priyayization: Path to Power? In: E. Locher-Scholten & A. Niehof (eds.) *Indonesian Women in Focus.* Leiden.

Doherty, Joseph
1975 Urban Places and Third World Development. *African Urban Notes*, Serie B, no. 2; pp. 1-17.

Drakakis-Smith, D. (ed.)
1986 *Urbanisation in the Developing World.* Beckenham: Croom Helm.

Dunham, D.
1991 *Agricultural Growth and Rural Industry; Some Reflections on the Rural Growth Linkages Debate.* The Hague: ISS Working Paper series, no. 114.

Dusseldorp, D.B.W.M. van
1971 *Planning of Service Centres in Rural Areas of Developing Countries.* Wageningen.Durand, J., W. Kandel, E.A. Parrado & D.S. Massey
1996 International Migration and Development in Mexican Communities. *Demography* 33; pp. 249-264.

Du Toit, B.M.
1990 People on the Move. Rural-Urban Migration with Special Reference to the Third World: Theoretical and Empirical Perspectives. *Human Organization* 49; pp. 305-319.

Economic and Social Commission for Asia and the Pacific (ESCAP)
1979 *Guidelines for Rural Centre Planning.* New York: United Nations.

ESCAP
1979a   *Economic and Social Commission for Asia and the Pacific*. Guidelines for Rural Centre Planning.
1979b   *Guidelines for Rural Centre Planning*. New York: United Nations
1990   *Guidelines for Rural Centre Planning*. Rural Industrialization.Organization Framework for RCP. New York: United Nations.

Evans, H.A.
1989   Farm Towns, Mining Towns and Rural Development in the Potosí Region, Bolivia. In: R. May (ed.), *The Urbanization Revolution*, New York, pp. 91-111.

Fair, T.J.D., G. Murdoch & H.M. Jones
1969   *Development in Swaziland: A Regional Analysis*. Johannesburg.

FAM
1976   Resumen cantonal - San Carlos. San José.

Fernandez, J.
1988   Inestabilidad económica con estabilidad política. San José: UCR.

Fields, C.O.N. et al.
1990   Labour Market Modelling and the Urban Informal Sector, Theory and Evidence. In: D. Turnham (ed.), *The Informal Sector Revisital*, Paris: OECD Development Center.

Firdausy, C.M.
1994   Urban Poverty in Indonesia: Trends; Issues and Policies. In: *Asian Development Review* vol. 12 no. 1; pp. 68-89.

Forbes, D.K.
1981   'Petty Commodity Production and Underdevelopment; The Case of Peddlers and Trishaw Riders in Ujung Pandang, Indonesia'. *Progress in Planning* vol. 16 (2); pp. 105-178.
1984   *The Geography of Underdevelopment*. London/Sydney: Croom Helm.

Fortuna, J.C. & S. Prates
1989   Informal Sector Versus Informalised Labour Relations in Uruguay. In: Portes, A., M. Castells & L.A. Benton (eds.), *The Informal Economy, Studies in Advanced and Less Developed Countries*. Baltimore/Londen: Johns Hopkins University Press, pp. 11-37.

Freeman, D.B. & G.B. Norcliffe
1984   Relation Between the Rural Non-Farm and Small Farm Sectors in Central Province, Kenya. *Tijdschrift voor Economische en Sociale Geografie* 75 (10); pp. 61-73.

Friedmann, J.
1966   *Regional Development Policy: A Case Study of Venezuela*. Cambridge, Mass: M.I.T. Press.

Friedmann, J. & M. Douglass
1978   *Agropolitan Development; Towards a New Strategy for Regional Planning in Asia*. In: Lo & Salih (eds.); pp. 163-192.

Friedmann, J. & F. Sullivan
1974   The Absorption of Labor in the Urban Economy: The Case of Developing Countries. *Economic Development and Cultural Change* 22; pp 385-413
1975   The Absorption of Labour in the Urban Economy: the Case of Developing Countries. In: Friedmann, J. & W. Alonso (eds.), *Regional Policy: Readings in Theory and Applications*, Cambridge (Mass.)/London: M.I.T Press, pp. 473-501.

Friedmann, J. & C. Weaver
1979   *Territory and Function; the Evolution of Regional Planning*. London: Arnold.

Frieling, H.D. von
1989   Das Konzept des informellen Sektors. Kritiek eines Entwicklungsidealismus. In: E.W. Schamp (ed.), *Der informelle Sektor. Geographische Perspektiven eines umstrittenen Konzepts*, Aachen: Alano Verlag/Edition Herodot, pp. 169-201.

Funnel, O.C.
1976   The Role of Small Service Centres in Regional and Rural Development; with Special Reference to East Africa. In: A. Gilbert (ed.), *Development Planning and Spatial Structure*. London: Arnold; pp. 77-112.

Gaile, G.L.
1992   Improving Rural-Urban Linkages Through Small-Town Market Based Development. *Third World Planning Review* 14, pp. 131-148.

Gallais, J.
1967   *Le Delta Intérieur du Niger, étude géographique régionale*. Dakar: IFAN.

Garnier, L.
1989   Crísis, desarrollo y democrácia en Costa Rica. In: E. Torres Rivas, ed., Costa Rica: crisís y desafios. San Jose: DEI, pp. 29-45.

Geertz, C.
1963   *Peddlers and Princes; Social Change and Economic Modernization in Two Indonesian Towns*. Chicago: Chicago University Press.
1965   The Social History of an Indonesian Town. Massachusets Institute of Technology Press, Cambridge (Massachusets).

Gelder, P. van
1985   *Werken onder de boom. Dynamiek in de informele sektor: de situatie in groot Paramaribo*, Dordrecht: Foris Publications & Cinnaminson (USA), 244 pp.

Gelderloos, H.
1984   *Het labyrinth van de armoede: een arbeidsmarktonderzoek in de NoordMexicaanse grensstad Ciudad Juarez*, Discussiestuk no. 27, Utrecht: Department of Geography, State University.

Gerold-Scheepers, T. & W.M.J. van Binsbergen
1978   Marxist and Neo-Marxist Approaches to Migration in Tropical Africa. *African Perspectives* 1. Leiden: Afrika Studiecentrum.

Gerry, C.
1978   'Petty Production and Capitalist Production in Dakar: The Crisis of the Self-Employed'. *World Development* vol. 6 no. 9/10; pp. 1147-1160.
1979   Small Scale Manufacturing and Repair in Dakar: a Survey of Market Relations Within the Urban Economy. In: Bromley, R. & Ch. Gerry (eds.), *Casual Workers and Poverty in Third World Cities*, Chichester: John Wiley & Sons, pp. 229-250.
1985   Wagers and Wage Working: Selling Gambling Opportunities in Cali, Colombia. In: Bromley, R. (ed.), *Planning for Small Enterprises in Third World Cities*. Oxford: Pergamon Press, pp. 155-169.

Gilbert, A. & J. Gugler
1984   *Cities, Poverty, and Development. Urbanisation in the Third World*, London/Worchester: Oxford University Press, pp. 65-81.
1992   *Cities, Poverty and Development: Urbanization in the Third World*. Oxford/New York: Oxford University Press (2nd edition).

Godebauer, H. & R. Trienes
1995   Child Labour. In: D.R. Harriss (ed.) *Prisoners of Progress. A Review of the Current Indonesian Labour Situation*. Meppel: INDOC Krips Repro; pp. 57-70.

Goldscheider, C. (ed.)
1983   *Urban Migrants in Developing Countries. Patterns and Problems of Adjustment*. Boulder: Westview Press.

Gooneratne, W. & R. Obudho (eds.)
1997   *Contemporary Issues in Regional Development Policy, Perspectives of Eastern and Southern Africa*. Aldershot: Ashgate Publications Ltd.

Gosses, A., K. Molenaar, Q. Sluijs & R. Teszler (eds.)
1989   *Small Enterprises, New Approaches*, The Hague: Ministry of Foreign Affairs, Directorate General International Cooperation.

Graaf, C. de & P. van der Male
1992   *Tribuana & Karangsari. Een onderzoek naar de ruraal-urbane relaties tussen twee droge landbouw dorpen en een kleinstedelijk verzorgingscentrum*. Doktoraal scriptie, Utrecht: Vakgroep SGO.

Grant Anderson, A.
1980   'The Rural Market in West-Java'. *Economic Development and Cultural Change* vol. 28 no. 4; pp. 753-777.

Hagen Koo
1981   Centre-Periphery Relations and Marginalization; Empirical Analysis of the Dependency Model of Inequality in Peripheral Nations. *Development & Change* 12, pp. 55-76.
Haggblade, S., P. Hazell & J. Brown
1984   Farm-Non Farm Linkages in Rural Sub-Saharan Africa. *World Development* 17, pp. 1171-1201.
Hansen, N.M.
1971   *Intermediate-Size Cities as Growth. Applications for Kentucky; the Piedmont Crescent; the Ozarks and Texas.* New York: Praeger Publishers.
1990   Impacts of Small- and Intermediate-Sized Cities on Population Distribution: Issues and Responses. *Regional Development Dialogue* 11; pp. 60-79.
Hardoy, J.E. & D. Satterthwaite (eds.)
1986   *Small and Intermediate Urban Centres: Their Role in Regional and National Development in the Third World.* London: Hodder and Stoughton.
Harfst, J.R.
1988   *Delicias para quien? Veranderingen in de productiestructuur en de invloed op migratiepatronen van en naar een middelgrote Noord-Mexicaanse stad: Cd. Delicias.* Utrecht: Vakgroep Sociale Geografie van de Ontwikkelingslanden (doctoraalscriptie).
Hariri Hadi
1984a   The Impact of Circular Migration on a Village Economy. In: *Bulletin of Indonesian Economic Studies* vol 25 no. 1; pp. 53-75.
1984b   'Kecamatan Towns Can Stem the Invasion of Large Cities'. In: *Prisma the Indonesian Indicator* vol. 13 no. 32; pp. 50-54.
Hardoy, J.E. & D. Satterthwaite
1986   *Small and Intermediate Centres: Their Role in National and Regional Development.* London: Hodder & Stoughton.
Harriss, B. & J. Harriss
1988   Generative on Parasitic Urbanism? Some Observations from the Recent History of a South Indian Market Town. *Journal of Development Studies* 20, no. 3, pp. 82-101.
Harriss, J.
1982   Character of an Urban Economy; 'Small Scale' Production and Labour Markets in Coimbatore, Parts I and II. *Economic and Political Weekly*, vol. 17, nrs. 23 and 24, pp. 945-954, 993-1002.
1985   Our Socialism and the Subsistence Engineer: the Role of Small Enterprises in the Engineering Industry of Coimbatore, South India. In: Bromley, R. (ed.), *Planning for Small Enterprises in Third World Cities*, Oxford: Pergamon Press, pp. 137-153.
Hart, J.K.
1971   *Informal Income Opportunities and Urban Employment in Ghana.* Paper presented at the conference on Urban Unemployment in Africa at the Institute of Development Studies: University of Sussex.
1973   Informal Income Opportunities and Urban Employment in Ghana. *Journal of Modern African Studies*, vol. 11, no. 1, pp. 61-89.
1974   Migration and the Opportunity Structure: a Ghanaian Case-Study. In: S. Amin, (ed.), *Modern Migrations in Western Africa*. London: Oxford University Press, pp. 321-342.
Harts-Broekhuis, E.J.A. & A. de Jong (eds.)
1985   *Investigation Socio-Economique de la ville de Djenné et ses Environs.* Rapport VII. Etudes approfondies du milieu rural. Bamako, Mali: Université d'Utrecht/Institut des Sciences Humaines.
1993   *Subsistence and Survival in the Sahel; Responses of Households and Enterprises to Deteriorating Conditions and Development Policy in the Mopti Region of Mali.* Dissertation, Utrecht: KNAG/NGS.
Harts-Broekhuis, E.J.A. & G.J. Tempelman
1983   *Kleine steden en regionale ontwikkeling; een wisselwerking? Een studie naar de sociaal-economische ontwikkeling van de steden Djenné en San (Mali). Discussiestukken van de Vakgroep Sociale Geografie van de Ontwikkelingslanden* nr. 19. Utrecht: Universiteit van Utrecht.
Hasibuan, S.
1987   *Small-Scale Industry Development in Indonesia.* Paper presented at the policy workshop on small-Scale

Industrialization. May-June 1987, Institute of Social Studies; The Hague: The Netherlands.
Herlaar, M. & M. Sonnema
1987   Het belang van de inkomsten uit kleinschalige aanvullende activiteiten voor rurale huishoudens in Meru District, Kenya. Unpublished Master Thesis. Utrecht: Universiteit van Utrecht.
Hetler, C.B.
1984   The Impact of Circular Migration on a Village Economy. *Bulletin of Indonesian Economic Studies* vol. 25 no. 1; pp. 53-75.
Hilhorst, Jos G.M.
1971   *Regional Planning; A Systems Approach*. Rotterdam: Rotterdam University Press.
Hinderink, J.
1983   *De rol van kleine steden in regionale ontwikkeling*. Paper Studiedag Sectie SGO van het KNAG. Discussiestukken van de Vakgroep Sociale Geografie van de Ontwikkelingslanden nr. 19. Utrecht: University of Utrecht.
1997   Small Towns, Big Dreams? in: Naerssen, T. van, M. Rutten & A. Zoomers (eds.), *The Diversity of Development; Essays in Honour of Jan Kleinpenning*. Assen: Van Gorcum, pp. 180-189.
Hinderink, J. & J.J. Sterkenburg
1978   *Anatomy of an African Town, a Socio-Economic Study of Cape Coast, Ghana*. Utrecht: Geographical Institute.
1987   *Agricultural Commercialization and Government Policy in Sub-Saharan Africa*. London: Kegan Paul.
Hinderink, J. & M.J. Titus
1988   Paradigms of Regional Development and the Role of Small Centres. *Development and Cultural Change* no. 19; pp. 401-425.
Hoenderdos, W.
1982   *Het arbeids- en produktiebestel in steden van ontwikkelingslanden en de 'urban poor'*. Thesis, Amsterdam: Department of Geography and Planning, Free University Amsterdam, unpublished.
Hoenderdos, W. & W. de Regt
1991   Migración y ciudad intermedia: Hidalgo del Parral, México. *Revista Interamericana de Planificación* 93; pp. 86-107.
Hofmann, M.
1986   The Informal Sector in an Intermediate City; A Case in Egypt. *Economic Development and Cultural Change*, no. 3, pp. 263-276.
Hofstede, G.
1983   *Culturele problemen voor Nederlandse managers in Indonesië*. Deventer: Twijnstra en Gudde International.
Holm, M.
1995   The Impact of Structural Adjustment on Intermediate Towns and Urban Migrants. An Example from Tanzania. In: Simon, D., W. van Spengen, C. Dixon & A. Närman (eds.). *Structurally Adjustment Africa; Poverty, Debt and Basic Needs*. London: Boulder.
Hopkins, N.S.
1979   The Small Centre in Rural Development; Kita (Mali) and Testour (Tunesia). *Africa*, vol. 49 (3), pp. 316-328.
Horst, F. v.d.
1992   *De arbeidsmarkt van Naranjo-stad*. Thesis, Utrecht: Department of Geography, State University, unpublished.
House, W.J.
1987   Labour Market Differentiation in a Developing Economy: an Example From Urban Juba, Southern Sudan, *World Development*, vol. 15, no.7, pp. 877-897.
Hugo, G.J.
1982   Circular Migration in Indonesia. *Population and Development Review* 8 (1); pp. 59-83.
1996   Urbanization in Indonesia: City and Countryside Linked. In: J. Gugler (ed.), *The Urban Transformation of the Developing World*. Oxford/New York: Oxford University Press, pp. 132-183.
Hugon, Ph., Nhu, A. le & A. Morice
1977   *La production marchande et l'emploi dans le secteur "informel" - le cas Africain*. Paris: Université I, Institut

d'Etude du Développement Economique et Social.

Huisman, H.
1994 Planning for Rural Development: Experiences and Alternatives. *Cases from Indonesia and Lesotho*. Utrecht: Koninklijk Nederlands Aardrijkskundig Genootschap/Geographical Institute Universty Utrecht Nederlandse Geografische Studies 180.

Huisman, H. & W. Stoffers
1988 Socio-Economic Conditions in Varying Settings within a District. A First Report on the Situation in Bantul District. *RRDP Research Report no 1*, Yogyakarta.
1990 Households, Resources and Production. A Second Report on the Situation in Bantul District. *RRDP Research Report no 2*, Yogyakarta.
1992 Spatial and Social Differentiation in the Use of Rural Community Services on Java. An Example from Bantul District, Special Province of Yogyakarta. *The Indonesian Journal of Geography* 23-25, pp. 87-184.

ILO
1972 *Employment; Incomes and Equality: a Strategy for Increasing Productive Employment in Kenya*. International Labour Office; Geneva.
1993 *World Labour Report*, Geneva: International Labour Office.

Ingram, D.R.
1971 The Concept of Accessibility, *Regional Studies* no. 5, pp. 101-107.

Jaarsma, A.
1994 Voorzien van voorzieningen? Sociaal-economische studie naar het voorzieningengebruik door de huishoudens uit het commerciële Hoogland Gebied van Kecamatan Kejajar; Midden Java. Utrecht; unpublished thesis.

Jamal, V.
1995 *Structural Adjustment and the Rural Labour Markets in Africa*. London: McMillan Press.

Jayasuriya, S.K. & R.T. Shand
1983 *Technical Change and Labour Absorption in Asian Agriculture; an Assessment*. Paper for the Conference on Off-Farm Employment in the Development of Rural Asia, Chiang Mai Thailand.

Johnson, E.A.J.
1970 *The Organization of Space in Developing Countries*. Cambridge, Mass.: Harvard University Press.
1974 *The Organization of Space in Developing Countries*. Cambridge.

Jones, G.W.
1984 Links Between Urbanization and Sectoral Shifts in Employment in Java. *Bulletin of Indonesian Economic Studies* vol. 20; pp. 120-155.

Jones, G.W. & C. Manning
1992 Chapter 11. Labour Force and Employment During the 1980s. In: A. Booth (ed.). *The Oil Boom and After*, pp. 363-410.

Jones, H.R.
1981 Chapter 8; Internal Migration. In: *A Population Geography*; pp. 199-250. London.

Jones, R.C.
1996 *Ambivalent Journey. US Migration and Economic Mobility in North-Central Mexico*. Tucson: University of Arizona Press.

Jong, A.A.
1983 Stedelijke groeiprocessen in West Afrika. Paper Studiedag Sectie SGO van het KNAG. *Discussiestukken Vakgroep Sociale Geografie van de Ontwikkelingslanden* nr. 19. Utrecht: Universiteit van Utrecht.

Jong, A.W. de & M. Ligthart
1986 *Small Town Productive Activities and Regional Development. The Case of Wonosobo; Central Java*. Utrecht; unpublished thesis.

Jong, W. de & F. van Steenbergen
1985 *Town and Hinterland in Central Java: the Banjarnegara Production Structure in Regional Perspective*. Unpublished M.A. Thesis. Department of Human Geography of Developing Countries. University of Utrecht.
1987 *Town and Hinterland in Central Java. The Banjarnegara Productionstructure in Regional Perspective*. Gadjah Mada University Press.

Joshi, H., H. Lubell & S. Mouly
1976   *Abidjan, Urbanization and Employment*. Geneve: International Labour Organization.
Kabwegyere, T.B.
1979   Small Urban Centres and the Growth Underdevelopment in Rural Kenya. *Africa* 49; pp. 308-315.
Kaihatu, H. & C. Ruitenbeek
1993   *Over leven in desa Kalimandi*. Utrecht; unpublished thesis.
Keene, B.
1980   Incursiones del Banco Mundial en Centroamerica. In: H. Assman, ed., El Banco Mundial: un caso de 'progressismo conservador'. San José: DEI, pp. 199-218.
Kievid, A.
1993   *Armoede; ongelijkheid en milieudegradatie. Een onderzoek naar twee hoogland landbouwsystemen in de bergen van Midden-Java*. Utrecht; unpublished thesis.
Klapwijk, M.
1989   *Purbalingga's Pushed and Pulled. Patterns of Labour Absorption in a Small Town in Central Java*. Diskussiestuk no. 42, Utrecht: Department of Geography, State University.
Kleijburg, J.P.
1988   *Verandering en arbeidsmobiliteit in Cd. Delicias*. Utrecht: Vakgroep Sociale Geografie van Ontwikkelingslanden (doctoraalscriptie).
Koentjaraningrat
1967   Tjelapar: A Village in South Central Java. In: Koentjaraningrat (ed.). *Villages in Indonesia*. New York.
Kragten, M. & J. Pieters
1991   *Desa's stuwend of gestuwd? Onderzoek naar de relatie van twee natte rijstdorpen met de kleine stad Wonosobo; Midden Java*. Utrecht; unpublished thesis.
Krings, Th.
1987   Surviving in the Periphery of the Town - the Living Conditions of Sahelian Refugees in Mopti (Republic of Mali). *Geo-Journal* 14, pp. 63-70.
Kubbinga, R.
1989   *Consumptiepatronen en het gebruik van voorzieningen in Wonosobo. Een huishoudonderzoek in een kleine stad op Java*. Indonesië. Utrecht; unpublished thesis.
Kuklinski, A. (ed.)
(1978) *Regional Policies in Nigeria, India and Brazil*. The Hague: Mouton.
Lanzetta de Pardo, M., G. Murillo C. & A. Trianas
1989   The Articulation of Formal and Informal Sectors in the Economy of Bogotá, Colombia. In: Portes, A., M. Castells & L.A. Benton (eds.), *The Informal Economy, Studies in Advanced and Less Developed Countries*. Baltimore/London: Johns Hopkins University Press, pp. 95-110.
Lebrun, D. & C. Gerry
1975   'Petty Producers and Capitalism'. *Review of African Political Economy* no. 3; pp. 20-32.
Lee, E.
1966   A Theory of Migration in: *Demography* 3, pp. 47-57.
1981   Basic Needs Strategies: a Frustrated Response to Development from Below? In: Stöhr & Taylor (eds.); pp. 107-122.
Lefeber, L.
1978   *Spatial Population Distribution; Urban and Rural Development*. Paper for the ECLA/CELADE Seminar on Population Redistribution.
Lewin, A.C.
1985   'The Dialectic of Dominance: Petty Production and Peripheral Capitalism'. In: Bromley R. (ed.). *Planning for Small Enterprises on Third World Cities*. Oxford: Pergamon Press; pp. 107-136.
Lier, P. van
1988   *De arbeidsmarktstruktuur an de arbeidsmogelijkheden in Hidalgo del Parral*. Thesis, Utrecht: Department of Geography, State University, unpublished.
Lin Lean Lim
1993   The Feminization of Labour in the Asia-Pacific Rim Countries: From Contributing to Economic Dynamism

to Bearing the Brunt of Structural Adjustments. In: Naohiro Ogawa G.W. Jones & J.G. Williamson (ed.). *Human Resources in Development Along the Asia-Pacific Rim*; pp. 175-209.

Lindert, P. van & O. Verkoren
1991 Ciudades intermediar y pequeñas, relaciones rural-urbanas y desarrollo regional en América Latina. *Revista Interamericanan de Planificación* 93, pp. 7-20.
1991a The Absorption Capacity of Small Towns in Latin America: Examples from Mexico and Bolivia. In: J.M.G. Kleinpenning ed. *The Incorporative Drive. Examples from Latin America*. Saarbrücken/Fort Lauderdale: Breitenbach; pp. 254-274.

Lindert, P. van & O. Verkoren (eds.)
1997 *Small Towns and Beyond. Rural Transformations and Small Urban Centres in Latin America*. Amsterdam: Thela Publishers.

Livingstone, Ian
1986 *Rural Development, Employment and Incomes in Kenya*. Aldershot: Gower.

Lo, Fu-Chen (ed.)
1981 Rural-Urban Relations and Regional Development. *UNCRD Regional Development Series* vol. 5; Nagoya.

Lo, Fu-Chen & K. Salih
1978 *Growth Pole Strategy and Regional Development Policy. Asian Experience and Alternative Approaches*. Oxford: Pergamon Press.
1981 Growth Poles, Agropolitan Development and Polarization Reversal. *The Debate and Search for Alternatives*. In Stöhr & Taylor (eds.); pp. 123-154.

Lowder, S.
1986 *Inside Third World Cities*. London: Croom Helm.

Luit. W. van der
1992 *Naranjo primero? Een onderzoek naar ruimtelijke mobiliteitsstromen vanuit en naar een kleine stad en haar ommeland in de Centrale Vallei van Costa Rica*. Utrecht: Vakgroep Sociale Geografie van de Ontwikkelingslanden (doctoraalscriptie).

Mabogunje, Akin L.
1978 Growth Poles and Growth Centres in the Regional Development of Nigeria. In A. Kuklinski (ed.); pp. 3-82.
1980 *The Development Process. A Spatial Perspective* London.

MacEwen Scott, A.
1979 Who are the Self-Employed? In R. Bromley & Ch. Berry (eds.), *Casual Work and Poverty in Third World Cities*. Chichester: John Wiley, pp. 87-104.

Mai, U.
1984 'Small Town Markets and the Urban Economy in Kabupaten Minahassa'. *Indonesia* no. 37; April 1984.

Manning, C.
1987 Rural Economic Change and Labour Mobility: A Case Study from West Java. *Bulletin of Indonesian Economic Studies* vol. 23; no. 3.

Manshard, Walther
1974 *Tropical Agriculture*. London & New York: Longman.
1977 *Die Städte des tropische Afrika*. Stuttgart: Bornträger.

Massey, D, R. Alarcón, J. Durand & H. González
1987 *Return to Aztlan. The Social Process of International Migration from Western Mexico*. Berkeley: University of California Press.

Mathur, O.P.
1982 *Small Cities and National Development*. Nagoya: United Nations Centre for Regional Development.

McGee, T.G.
1971 'Catalysts or Cancers? The Role of Cities in Asian cities'. In: L. Jakobson & V. Prakash (eds.). *Urbanization and National Development*. Beverly Hills: Sage; pp. 159-182.
1976 The Persistence of the Proto-Proletariat; Occupational Structures and Planning for the Future of Third World Cities. *Progress in Geography* 9, pp,. 3-38.
1979 'Conservation and Dissolution in the Third World City: the Shantytown as an Element of Conservation'. *Development and Change* vol. 10; pp. 1-22.

1981   Labour Mobility in Fragmented Labour Markets; Rural-Urban Linkages and Regional Development in Asia. In: Fu-Chen Loo (ed.), *Rural-Urban Relations and Regional Development*, Nagoya, UNCRD, pp.

McGee, T.G. & R. Chandra
1984   'Comment'. *Regional Development Dialogue*; vol. 5 no. 2; pp. 180-183.

Meilink, H.A.
1978   Some Economic Interpretations of Migration. *African Perspectives* 1. Leiden: Afrika Studiecentrum.
1983   *The African City*, London: Methuen.

Meilink, H.A. & W.M.J. van Binsbergen
1978   Migration and the Transformation of Modern African Society. Introduction. *African Perspectives* (1); pp. 7-19.

Mellor, J.W.
1990   Agriculture on the Road to Industrialization. In: Eichle, C.K. & J.M. Saatz (eds.), *Agricultural Development in the Third World*. Baltimore: Johns Hopkins Univesrity Press, pp. 70-88.

Menke, J.
1980   Marginalisering en werkloosheid in Groot-Paramaribo. *Sociaal Economisch Tijdschrift Suriname*, vol. 1, no. 1, pp. 57-85.

Misra, R.P. & K.V. Sundaram
1978   *Growth Foci as Instruments of Modernization in India*. In Kuklinski (ed.); pp. 98-113.

Missen, G.J.
1972   *Viewpoint on Indonesia; A Geographical Perspective*. Melbourne: Nelson.

Moir, H.V.J.
1977   *Labor Force and Labor Utilization on Selected Areas on Java*. Jakarta: Leknas-Lipi.

Morris, A.
1981   *Latin America; Economic Development and Regional Differentiation*. London: Hutchinson.

Mosely, M.J.
1979   *Accessibility, the Rural Challenge*. London:Methuen.

Moser, C.O.N.
1978   'Informal Sector or Petty Commodity Production: Dualism or Dependence in Urban Development?'. *World Development* vol. 6 no. 9/10; pp. 1041-1064.
1984   *The Role of the Informal Sector in Small and Intermediate-Sized Cities*. UNCRD Paper, Nagoya.

Moser, C.O.N., et al.
1984   The Informal Sector Reworked: Viability and Vulnerability in Urban Development. *Regional Development Dialogue*, vol. 5, no. 2, pp. 135-178.
1993   *Urban Poverty in the Context of Structural Adjustment; Recent Evidence and Policy Response*, Washington D.C.: World Bank.

Mosher, A.T.
1969   *Creating a Progressive Rural Structure*. New York: Agricultural Development Council.

Mourmans, W.
1981   Het dualisme begrip in de analyse van ontwikkeling. In: *CASP* serie werkdocumenten no. 2; pp. 34-66.

Muñoz, H., O. de Oliveira & C. Stern
1982   Selected Studies on the Dynamics Patterns and Consequences of Migration. *I. Mexico City: Industrialization Migration and the Labour Force 1930-1970* (*Reports and Papers in the Social Science* 46). France: UNESCO.

NUDS
1985   *Laporan akhir NUDS* (Proyek Strategi Nasional Pengembangan Perkotaan). DPU-UNDP Jakarta.

Nuhn, H.
1989   *Dezentralisierungspolitik und Raumbezogene staatliche Entwicklungsplanung in Costa Rica 1960-1988*. Hamburg: Institut fur Geographie und Wirtschaftsgeographie der Universitat Hamburg, unpublished.

OECD
1988   *The Sahel Facing the Future; Increasing Dependence or Structural Transformation?* Paris: OECD

Oliveira, O. de & B. Roberts
1996   Urban Development and Social Inequality in Latin America. In: J. Gugler ed. *The Urban Transformation of*

the Developing World. Oxford: Oxford University Press; pp. 253-314.

Oosterhout, A.T.H. van
1985 *Region and Reign; Regionale ongelijkheid en regionaal beleid in de Filippijnen.* Nijmegen: Nijmeegse Geografische Cahiers 27.

Ottens, B.J.
1988 *Purbalingga als verzorgingscentrum. Het verzorgingsapparaat van een kleine stad in Midden Java in een regionaal perspektief.* Utrecht unpublished thesis.

Pedersen, P.O.
1990 The Role of Small Rural Towns in Development, in: J. Baker (ed.), *Small Town Africa; Strikes in Rural-Urban Interaction.* Uppsala: The Scandinavian Institute of African Studies, Seminar Proceedings 23, pp. 89-107.
1991 The Role of Small and Intermediate Urban Centres in Planning in Africa. *African Urban Quarterly* 6, pp. 171-243.
1997 *Small African Towns: Between Rural Networks and Urban Hierarchies.* Aldershot: Ashgate Publications Ltd.

Peluso, N.L.
1984 *Occupational Mobility and the Economic Role of Women.* Gadjah Mada University Yogyakarta.

Pemda Bantul
1991 *Bantul Dalam Angka.*
1995 *Bantul Dalam Angka.*

Pernia, E.M. & D.N. Wilson
1989 Education and Labor Markets in Indonesia. A Sector Survey. *Asian Development Bank Economic Staff Paper* no. 45. Manila: Asian Development Bank. Population of Jawa Tengah. Results of the 1980 Population Census; pp. 93-94. Jakarta: BPS.

Post, C. van der
1988 *Migrants and Migrant Labour Absorption in Large and Small Centres in Swaziland; A Comparative Study of the Towns Manzini and Vhlangano.* Dissertation, Utrecht: KNAG/NGS.

Preston, D.
1996 People on the Move: Migrations Past and Present. In: D. Preston (ed) *Latin American Development. Geographical Perspectives* (Second Edition). London: Longman; pp. 165-187.

Quijano Obregon, A.
1974 The Marginal Pole of the Economy and the Marginalised Labour Force. *Economy and Society*, no. 3, pp. 393-428.

Ramos Pioquinto, D.
1991 Migración y cambios socioeconómicos en la comunidad de Zoogocho Oaxaca. *Estudios Demográficos y Urbanos* 6-2, pp. 313-345.

Regt, W. de
1987 *Een middelgrote stad in beweging, patronen en invloeden van migratie naar en uit Hidalgo del Parral Chihuahua (Mexico).* Utrecht: Vakgroep Sociale Geografie van de Ontwikkelingslanden (doctoraalscriptie).

Rees, M.W & A.D. Murphy; E.W. Morris & M. Winter
1991 Migrants To and in Oaxaca City. *Urban Anthropology* 20, pp. 15-29.

Republique du Mali
1988 *Recensement National.*

Rietveld, P.
1986 'Non-Agricultural Activities and Income Distribution in Rural Java'. *Bulletin of Indonesian Economic Studies* 22 (3); pp. 106-117.

RIM
1987 *Un refuge dans le Sahel. Le cheptel, et les systèmes de production dans la cinquième Région au Mali.* Ressource Inventory and Management Ltd, Compendium House Jersey.

Roberts, B.R.
1976 The Provincial Urban System and the Process of Dependency. In: A. Portes & H.L. Browning eds. *Current Perspectives in Latin American Urban Research.* Austin: University of Texas Press; pp. 99-131.
1978 *Cities of Peasants; the Political Economy of Urbanization in the Third World.* London: Edward Arnold.

1995 *The Making of Citizens; Cities of Peasants Revisited*. London: Edward Arnold.

Roberts, P.R.
1989 Employment Structure, Life Cycle, and Life Changes: Formal and Informal Sectors in Guadalajara. In: Portes, A., M. Castells & L.A. Benton (eds.) - *The Informal Economy, Studies in Advanced and Less Developed Countries*, Baltimore/London: Johns Hopkins University Press, pp. 41-59.

Rodriguez, A.G. & S.M. Smith
1994 A Comparison of Urban Rural and Farm Poverty in Costa Rica. *World Development* 22; pp. 381-397.

Romein, A.
1995a *Labour Markets and Migrant Absorption in Small Towns; The Case of Northern Costa Rica*. Dissertation, Utrecht: KNAG/NGS.
1995b *Urban Labour Markets and Migrant Absorption in Small Towns: The Case of Northern Costa Rica*. Utrecht: Koninklijk Nederlands Aardrijkskundig Genootschap / Faculteit Ruimtelijke Wetenschappen.

Romein, A. & J. Schuurman
1990 Regional Strategy in Costa Rica and its Impact on the Northern Region. In: D. Simon (ed.) *Third World Regional Development; A Reappraisal*. London: Chapman, pp. 94-108.

Rondinelli, D.A.
1983 *Secondary Cities in Developing Countries: Policies for Diffusing Urbanization*. Beverly Hills: Sage.
1985 *Applied Methods of Regional Analysis. The Spatial Dimensions of Development Policy*. Boulder/London: Westview Press.
1988 Market Town and Agriculture in Africa; The Role of Small Urban Centres in Economic Development. *African Urban Quarterly* 3, no. 1/1, pp. 3-10.
1991 Asian Urban Development Policies in the 1990s; From Growth Control to Urban Diffusion. *World Development* 19, no. 7, pp. 791-803.

Rondinelli, D.A. & H. Evans
1983 Integrated Regional Development Planning: Linking Urban Centres and Rural Areas in Bolivia. *World Development* Vol. 11 (1); pp. 31-53.

Rondinelli, D.A. & K. Ruddle
1978 *Urbanization and Rural Development. A Spatial Policy for Equitable Growth*. New York: Praeger.

Root, B.D & F. de Jong
1991 Family Migration in a Developing Country. *Population Studies* 45, pp. 221-233.

Rushton, G.
1984 Use of Location-Allocation Models for Improving the Geographical Accessibility of Rural Services in Developing Countries. *International Regional Science Review* 9, pp. 217-240.

Sadoko, J.
1989 *Growth of Small and Intermediate Cities and Manufacturing Development in West-Java: A Research Paper*. The Hague: ISS Working Paper Series, no. 68.

Sanchez, C.F., H. Palmiero & F. Fernando
1981 The Informal and Quasi-Formal Sector in Córdoba. In: S.V. Sethuraman, (ed.), *The Urban Informal Sector in Developing Countries: Employment, Poverty and Environment*. Geneva: ILO, pp. 144-158.

Santos, M.
1975 *L'espace partagé. Les deux circuits de l'économie urbaine des pays sous-développés*. Paris: Génin,
1979 *The Shared Space, the Two Circuits of the Urban Economy in Underdeveloped Countries*. London: Methuen.

Schamp, E.W.
1989 Was ist informell? Eine Einführung aus Sicht der Geographen. In: E.W. Schamp (ed.), *Der informelle Sektor. Geographische Perspektiven eines umstrittenen Konzepts*, Aachen: Alano Verlag/edition Herodot, pp. 7-33.

Schatzberg, M.
1979 'Islands of Privilege: Small Cities in Africa and the Dynamics of Class Formation'. *Urban Anthropology* vol. 8 (2); pp. 173-190.

Schillemans, P.I. & A.G. Verbeek
1989 *Op zoek naar werk. Een klein-stedelijk arbeidsmarktonderzoek in Banjarnegara* Midden-Java. Utrecht; unpublished thesis.

Schippers, H. & H. Thorborg
1994   *Bobotsari: Dynamisch? Een onderzoek naar marktrelaties tussen een kleine stad en haar rurale ommeland; Midden Java-Indonesië*. Utrecht; unpublished thesis.

Schmitz, H.
1982   Growth Constraints on Small-Scale Manufacturing in Developing Countries: a Critical Review. *World Development* 10, pp. 429-450.

Schumpeter, J.A.
1942   *Capitalism, Socialism and Democracy*. New York: Harper.

Sethuraman, S.V.
1981   *The Urban Informal Sector in Developing Countries: Employment, Poverty and Environment*, Geneva: ILO, pp. 1-47, 171-208.

Siamwalla, A.
1978   Farmers and Middlemen: Aspects of Agricultural Marketing in Thailand. *Economic Bulletin for Asia and the Pacific* 29, pp. 38-50.

Sibbing, G.
1984   *Migration Employment and Development: On the Role of Small Towns of Minas Gerais Brazil*. In: D. Kammeier & P. Swan, pp. 478-488.

Skoruppa, S.
1982   Diferenciación agraria y relaciones ciudad-campo en areas perifericas de Costa Rica. El caso de las subregiones Ciudad Quesada/San Carlos y Liberia/Guanacaste. In: M. Morales Alvarez & G. Sandner, eds., *Regiones perifericas y ciudades intermedias en Costa Rica*, San Jose: EUNED, pp.255-316.

Slater, D.
1986   *Capitalism and Urbanisation at the Periphery: Problems of Interpretation and Analysis with Reference to Latin America* in: Drakakis-Smith (ed.).

Smith, S.J.
1988   Kleinschalige industrie in Latijns Amerika: een studie van de ontwikkelingsmogelijkheden van de 'informele' kleding en textielnijverheid in Aguascalientes, Mexico. *Nederlandse Geografische Studies*, no. 70, Amsterdam: Centrum voor Educatieve Geografie, Vrije Universiteit, 413 pp.

Soegiyoko, S. & B.T. Sugiyanto
1976   Urban Areas in Indonesia as Development Catalyst. *Prisma* 2-3; pp. 52-62.

Soto, H. de
1989   *The Other Path, the Invisible Revolution in the Third World*. London/New York: Methuen.

Southall, A. (ed.)
1979   *Small Urban Centres in Rural Development in Africa*. Winconsin: Madison.
1988   Small Towns in Africa Revisited. *African Studies Review* 31, no. 3, pp.

Stam, F.
1991   *Bienvenidos? De migratie naar en vanuit de kleine stad Nuevo Casas Grandes in Chihuahua, Mexico*. Utrecht: Vakgroep Sociale Geografie van de Ontwikkelingslanden (doctoraalscriptie).

Stearman, A.M.
1985   *Camba and Kolla Migration and Development in Santa Cruz, Bolivia*. Orlando: University of Central Florida Press.

Stöhr, W.B.
1975   Regional Development. Experiences and Prospects in Latin America. *Series Regional Planning* vol. 3. Den Haag: Mouton.
1981   Development from Below. The Bottom-Up and Periphery Inward Development Paradigm. In Walter Stöhr & D.R.F. Taylor (eds.); pp. 39-71.

Stöhr, W.B. & D.R.F. Taylor (eds.)
1981   *Development from Above or Below? The Dialectics of Regional Planning in Developing Countries*. Chichester: Wiley.

Stonich, S.C.
1991   Rural Families and Income from Migration: Honduran Households in the World Economy. *Journal of Latin American Studies* 23, pp. 131-161.

Swaziland Government
1968 Report on the 1966 Swaziland Population Census. Mbabane.
1972 Second National Development Plan; 1973-1977. Mbabane.
1980 Report on the 1976 Swaziland Population Census. Mbabane.
n.d. a) Third National Development Plan; 1978/1979-1982/1983. Mbabane.
n.d. b) Fourth National Development Plan; 1983/1984-1987/1988. Mbabane.
1986 National Physical Development Plan (unpublished draft).

Taylor, D.R.F.
1974 The Role of the Smaller Urban Place in Development. A Case Study from Kenya. *Proceedings of the Commission of Regional Aspects of the Development Process*. London: University of Ontaria; pp. 615-638.

Taylor, D.R.F. & R.A. Obudho (eds.)
1979 *The Spatial Structure of Development. A Case Study of Kenya*. Boulder, Col.: Westview Press.

Teeffelen, P.B.M. van et al.
1981 *Investigation Socio-Economique de la Ville de Djenné et ses Environs*, cycle III. Rapport préliminaire: la planification régionale, les services à Djenné, la structure socio-économique des villages alentours. Bamako, Mali: Institut des Sciences Humaines.

Teeffelen, P.B.M. van
1992 *Dienstencentra en rurale ontwikkeling (Service Centres and Rural Development). Een onderzoek naar het aanbod en gebruik van overheidsdiensten in Mali, West-Afrika*. Dissertation, *Netherlands Geographical Studies* 138, Utrecht: KNAG.

Teilhet-Waldorf, S. & W.H. Waldorf
1983 Earnings of Self-Employed in an Informal Sector: A Case Study of Bangkok. *Economic Development and Cultural Change* vol. 31 (3); pp. 587-607.

Terhorst, R. & E. van Maanen
1992 *Jonggolsari Wonosobo. Een onderzoek naar de ruraal-urbane relaties tussen een droog landbouwdorp en een kleine stad Midden-Java*. Utrecht; unpublished thesis.

Titus, M.J.
1982 Social-Spatial Consequences of the Integration of the Serayu Region (Central Java) Into the Colonial System. In: O. v.d. Muyzenberg et al (eds.) *Focus on the Region in Asia*. Rotterdam: Casp Series.
1985 *Urbanisation Integration and Demographic Response in Jakarta. An Empirical Search for the Urban Role in Modernisation, 1961-1976*, Netherlands Geographical Studies 3, Amsterdam/Utrecht: KNAG.
1989 *Small Town Production Relations and Regional Functions in Central Java - Indonesia*. Unpublished Paper; Presented at the K.I.T.L.V. Workshop 10-15 September in Leiden.
1991 A Structural Model for Analyzing Production Relations and Regional Functions of Small Towns in Central Java. *Tijdschrift voor Economische en Sociale Geografie* 82, pp. 266-281.

Titus, M.J., W. de Jong & F. van Steenbergen
1986 Explorations Into the Economic Structure and Role of Small Urban Cities in the Serayu Valley Region, Central Java. In: P.J.M. Nas (ed.), *The Indonesian City: Studies in Urban Development and Planning*. Dordrecht: Cinnaminson, pp. 250-274.

Titus, M.J. & A.A. van der Wouden
1992 Manufacturing Activities in Four Towns in Central Java: An Assessment of Their Development Potential. *Singapore Journal of Tropical Geography* 13, pp. 130-151.

Titus, M.J., A. van der Wouden & M. Kragten
1994 Exploring Regional Aspects of Rural Development and Rural Diversification in Java. In: Harts-Broekhuis, A. & O. Verkoren (eds.), *No Easy Way Out. Essays on Third World Development in Honour of Jan Hinderink*. Utrecht: KNAG INES, pp. 291-307.

Tokman, V.E.
1978 An Exploration Into the Nature of Informal-Formal Sector Relationships. *World Development*, vol. 6, no. 9/10, pp. 1065-1075.

Trager, L.
1988 *The City Connection: Migration and Family Interdependence in the Philippines*. Ann Arbor: University of Michigan Press.

Turner, J.C.
1968   Housing Priorities. Settlement Patterns and Urban Development in Modernizing Countries in: *Journal of the American Institute of Planners* no. 36; pp. 354-363.

United Nations
1985   *Demographic Yearbook 1983*. New York.

United Nations Organisation (UNO)
1985   *Costa Rica. Economic Survey of Latin America and the Caribbean 1983*, vol.I, Santiago de Chile, pp. 209-234.

Uribe-Echevarría, F.
1991   Beyond the Informal Sector. Labour Absorption in Latin American Urban Economies, the Case of Colombia, *ISS Working Papers* no. 111, The Hague: Institute of Social Studies, pp. 21-60.

Velázquez Gutiérrez, L.A. & J. Arroyo Alejandro
1992   La transición de los patrones migratorios y las ciudades medias. *Estudios Demográficos y Urbanos* 7-2/3, pp. 555-572.

Verduzco, G
1984   Crecimiento urbano y desarrollo regional: el caso de Zamora Michoacán. *Revista Interamericana de Planificación*, pp. 67-80.

Verhoog, M. & M. van Rijn
1989   *Kleinstedelijke produktiestructuur en regionale ontwikkeling: De stad Purbalingga, Midden Java*, Unpublished Thesis, Utrecht.

Vletter, F. de
1982   Labour Migration in Swaziland: Recent Trends and Implications in: *South African Labour Bulletin* vol 7, no. 6. Durban.

Volbeda, S.
1997   *A Comparison Between Pioneer Towns and Rural Service Centres in the Amazon Region of Brazil*. In: Lindert, P. van & O. Verkoren (eds.), pp. 15-30.
1984   *Pionierssteden in het oerwoud. Stedelijke ontwikkelingen aan een agrarisch kolonisatiefront in het Braziliaanse Amazonegebied*. Dissertatie Nijmegen.

Vos, S. de & K. Richter
1988   Household Headship among Unmarried Mothers in Six Latin American Countries. *International Journal of Comparative Sociology* 29, pp. 214-230.

Vries, F. de
1985   *Migratie naar Ciudad Chihuahua: een verklaring vanuit de agrarische structuur van de staat*. Utrecht: Vakgroep Sociale Geografie van de Ontwikkelingslanden (doctoraalscriptie).

Waal, H.C.J.M. van der & A.A. van der Wouden
1988   *Dringen op de arbeidsmarkt. Een onderzoek naar de werkgelegenheidssituatie in de kleine stad Wonosobo - Indonesië*. Utrecht; Unpublished Thesis.

Wanmali, S.
1996   *Access to Rural Service and Economic Development in Communal Areas of Zimbabwe*. In: Gooneratne, W. & R.A. Obudho (eds.), pp. 215-229.

Wanmali, S. & Y. Isloon
1995   Rural Services; Rural Infrastructure and Regional Development in India, *Geographical Journal* 161, pp. 149-165.

Weaver, C.
1981   *Development Theory and the Regional Question; a Critique of Spatial Planning and its Detractors*. In: Stöhr & Taylor (eds.); pp. 73-106.

Weyland, H.J.W.
1984   Rural Industry. Sign of Poverty or Progress? *Collaborative Paper* nr. 6. Salatiga: Satya Wacana University.

White, B.
1976   Population, Involution and Employment in Rural Java. *Development and Change* 7; pp. 267-290.
1991   *In the Shadow of Agriculture; Economic Diversification and Agrarian Change in Java, 1900-1991*. The Hague: ISS Working Paper no. 96.

White, B. & H. Makali
1979   *Some Priliminary Notes from the Agro-Economic Survey*. The Hague: ISS.
White, P.E. & R.I. Woods
1980   Spatial Patterns of Migration Flows. In: P.E. White & R.I. Woods eds. *The Geographical Impact of Migration*. London: Longman; pp. 21-41.
Whittington, G.
1970   Towards Urban Development in Swaziland in: *Erdkunde* vol. xxiv, Heft 1.
Wilhelmy, H. & A. Borsdorf
1984   *Die Städten Sudamerikas*, teil 1 und 2, Berlin/Stuttgart: Gebrüder Bornträger.
World Bank
1979   *Urban Growth and Economic Development in the Sahel*. Washington: World Bank, Staff Working Paper 315
Woltman, O.
1990   *Gateway To an Additional Buck for the Poorest Mexican. Migratory Flows To and from Ojinaga*. Utrecht: Vakgroep Sociale Geografie van de Ontwikkelingslanden (doctoraalscriptie).
Wong, S.T.
1984   'Comparison of the Economic Impacts of Six Growth Centres on Their Surrounding Rural Areas in Asia'. In: *Environment and Planning* vol. 16 (1), pp. 81-94.
Woods, R.
1982   *Theoretical Population Geography*. London: Longman.
Wouden, A.A. van der
1991   *The Employment Structure and the Regional Functioning of Three Small Towns in Central Java*. Paper presented at a SOREGIO/ISS Symposium, unpublished, 29 pp.
1997   *Three Small Towns in Central Java; A Comparative Study of Their Economic Structure and Regional Importance*. Dissertation, Utrecht: KNAG/NGS.
Zabin, C. & S. Hughes
1995   Economic Integration and Labor Flows: Stage Migration in Farm Labor Markets in Mexico and the United States. *IMR*; pp. 395-422.
Zelinsky, W.
1971   The Hypothesis of the Mobility Transition in: *Geographical Review* 61; pp. 219-249.
Zoomers, A.
1997   Titling Land in Bolivia. Searching for a Redefinition of Tenure Regimes. In: Naerssen, T. van, M. Rutten & A. Zoomers (eds.), *The Diversity of Development*. Assen: Van Gorcum, pp. 59-70.
Zoomers, E.B.
1983   *Ruimtelijke dynamiek in een afhankelijke regio: oorzaken en gevolgen van migratie naar Ciudad Juárez Chihuahua Mexico*. Utrecht: Vakgroep Sociale Geografie van de Ontwikkelingslanden (doctoraalscriptie).
1984   Rurale dienstverlening als deel van regionaal ontwikkelingsbeleid: overheidsvoorzieningen en rurale ontwikkelling in de Cercle Mopti, Mali. Utrecht: Universiteit Utrecht, Geografisch Instituut, Discussiestukken van de Vakgroep SGO, 28.

Appendices

To Chapter Six:

Annex 6.1: Variables used for a typology of enterprises and institutions in Cd. Quesada

Variable	Indicator	Value 0	Value 1
Technological level	Use of modern equipment	Use of < 3 of the following: car, pick-up, lorry, electric and/or motorised machines, telephone, electric typewriter, personal computer	Use of > = 3
Organisation of production	Activities of owner/manager	Owner/manager performs a mix of blue and white-collar activities	Owner/manager performs only white-collar activities
Scale of operation	Size of monthly turnover or budget	< c 250,000 a)	> = c 250,000
Economic and technical conduct of business	Insurance of premises, vehicles or employees	No insurance	One or more types of insurances
	Bookkeeping	Not detailed or not at all	Detailed
	Access to source(s) of formal capital	No access	Access to one or more sources
	Use of bank account	None or not separate from household account	Separate from household account

a) US $ = c 90

Annex 6.2: The segmentation of the production structure of Cd. Quesada, 1990

	H	M	L
Services			
Hotel, catering & recreation		*	
Personal services			*
Economic and technical services & private health care			*
Agricultural services	*		
Financial services	*		
Governmmnent, public education & public ealth care	*		
Public utilities & passenger transport	*		
Cargo transport			
Socio-cultural services	*		
Trade			
Pulperias, tramos, butchers, bakeries & homebound sales			*
Supermarkets & wholesale of foodstuffs	*		
Garment, textile, shoes & sporting goods		*	
Transport equipment	*		
Agricultural inputs	*		
Electric domestic utensils, furniture & drugstores	*		
Iron monger & building materials		*	
Timber		*	
Other shops		*	
Street trade			*
Industry			
Sawmills	*		
Timber processing		*	
Bulk production of coffee, sugar and milk powder	*		
Construction			*
Empacadoras			*
Domestic food preparation (*Tacos*, icecreams etc.)			*
Leather processing			*
Tailoring			*
Other		*	
Repair			
Mechanical workshops			*
Domestic utensils			*
Shoes, watches & trinkets			*
Others			*

Annex 6.3: Characteristics of jobs in the quality of employment index

Employment status	Certainty of job	Part-/full-time job
0 = unpaid family-labourer	0 = irregular/temporal	0 = part-time
0 = self-employed worker	0 = seasonal	
1 = wage worker	1 = pemanent/fixed	1 = full-time
1 = manager/entrepreneur		

To Chapter Eight:

Annex 8.1: Delineation of the mode of production and form characteristics applied in the matrix model

Mode of production characteristics	
Capitalist mode	**Non-capitalist mode**
* Head of enterprise is engaged with sales and organization only	* Head of enterprise is engaged in all lines of work
* Wage workers employed	* No wage workers
* No family labour employed	* Family labour employed
* Value of capital goods exceeds Rp. 250,000 per person engaged	* Value of capital goods does not surpass Rp. 250,000 per person engaged
* The use of formal (bank) credits	* Private capital or informal credits used
* Profits are reinvested in the enterprise	* Profits are used in first instance for household consumption needs

Form characteristics	
Corporate sector	**Petty commodity sector**
* Monthly turnover - Rp. 5,000,000, or	* Monthly turnover < 5,000,000, or
* Value of capital goods > Rp. 5,000,000, or	* Volume of capital goods < Rp. 5,000,000, or
* Ten or more employees on the pay roll, or	* Less than 10 employees on the pay roll, or
* Value of real estate > Rp. 5,000,000, or rent real estate > Rp. 10,000 per month	* Value of real estate < Rp. 5,000,000, or rent real estate < Rp. 10,000 per month
* Registration at kabupaten (district) office	* Not registered or at most registered at the market office
* Use of modern machinery and/or motorized transport equipment	* Use of simple tools or households appliances at most
* Use of modern bookkeeping	* Very simple bookkeeping or none at all

Annex 8.2: Minimum necessary monthly income for the three district capitals, in 1986 and 1988

Minimum income level for:	Banjarnegara (a)	Purbalingga (b)	Wonosobo (c)
One person	Rp. 13.920	Rp. 9,450	Rp. 11,550
One person and average number of dependants	Rp. 37,584	Rp. 26,460	Rp. 25,410
Average-sized household (d)	Rp. 79,344	Rp. 54,810	Rp. 68,145

a) The calculation for Banjarnegara is based on the average price of rice in 1988 as calculated from table IX.4.1. (Banjarnegara Dalam Angka 1988).
b) The calculation for Purbalingga is based on the average price of rice mentioned in table 9.4.1 (central Java in figures, 1988; for 1986 for Purwokerto).
c) The average price of rice in Magelang in 1986, as mentioned in table 9.4.1 (Central Java in figures, 1988) has been used in the calculation for Wonosobo.
d) An average-sized household consist of 5.7 persons, 5.8 and 5.9 persons in Banjarnegara, Purbalingga and Wonosobo, respectively.

Annex 8.3: Age distribution of the sample populations (%)

Age category	Banjarnegara		Purbalingga		Wonosobo	
	Males N=587	Females N=593	Males N=645	Females N=685	Males N=866	Females N=846
0-4	9	11	9	10	8	7
5-9	13	10	16	15	11	10
10-14	14	11	15	13	14	12
15-19	12	11	15	12	15	14
20-24	10	11	6	8	10	9
25-29	9	9	7	8	6	7
30-34	6	8	6	7	6	8
35-44	10	9	10	9	6	9
45-54	6	9	8	9	10	13
55-64	7	8	5	8	7	6
65+	4	3	3	2	7	5

Source: Household survey

To Chapter Nine:

Annex 9.1: Weighed scores for the various groups of services for all settlements, Bantul District, 1990

Settlement*	Subdistrict	Population (688,195)	Community services	Production services	Commercial service	Total
Regional town						
Bantul	Bantul	13,289	13	9	36	58
District towns						
Bangunharjo	Sewon	15,232	11	9	25	45
Donotirto	Kretek	8,837	12	5	26	43
Karangtalun	Imogiri	2,776	5	10	26	41
Srimulyo	Piyungan	13,382	8	11	21	40
Ngestiharjo	Kasihan	19,073	8	6	25	39
Locality towns						
Panggungharjo	Sewon	18,640	12	4	19	35
Srihardono	Pundong	11,426	8	8	19	35
Argosari	Sedayu	7,802	9	7	17	33
Baturetno	Banguntapan	7,873	11	9	13	33
Tirtonirmolo	Kasihan	15,438	9	8	16	33
Wijirejo	Pandak	9,607	11	9	13	33
Argomulyo	Sedayu	10,879	10	9	13	32
Imogiri	Imogiri	3,432	4	4	24	32
Pleret	Pleret	9,107	11	10	11	32
Wonokromo	Pleret	8,556	9	5	17	31
Rural villages						
'High'						
Banguntapan	Banguntapan	20,921	7	3	19	29
Trimurti	Srandakan	15,820	10	6	13	29
Wukirsari	Imogiri	13,052	8	5	16	29
Palbapang	Bantul	11,958	9	5	14	28
Timbulharjo	Sewon	15,326	8	7	13	28
Ringinharjo	Bantul	6,337	7	2	18	27
Sumbermulyo	Bambanglipuro	14,834	7	8	12	27
Gadingsari	Sanden	10,657	7	10	9	26
Murtigading	Sanden	8,154	10	2	14	26
Pendowoharjo	Sewon	15,254	6	5	15	26
Panjangrejo	Pundong	9,217	6	3	16	25
Argodadi	Sedayu	9,475	4	5	15	24
Argorejo	Sedayu	7,987	5	5	14	24
Patalan	Jetis	11,090	5	6	13	24
Jagalan	Banguntapan	3,163	7	2	14	23
Sidomulyo	Bambanglipuro	2,516	10	3	10	23
Sumberagung	Jetis	10,478	6	6	11	23
Trimulyo	Jetis	12,058	7	6	10	23
Poncosari	Srandakan	11,923	8	3	11	22
Sabdodadi	Bantul	5,215	8	3	11	22
Tirtomulyo	Kretek	6,762	7	4	11	22
Tirtosari	Kretek	4,127	6	2	14	22
Sitimulyo	Piyungan	10,398	5	3	13	21

Annex 9.1 (continued)

Settlement*	Subdistrict	Population (688,195)	Community services	Production services	Commercial service	Total
Rural villages						
'Low'						
Trirengo	Bantul	14,643	10	1	9	20
Gilangharjo	Pandak	13,950	4	5	10	19
Srimartani	Piyungan	10,707	8	4	7	19
Tirtoharjo	Kretek	2,834	4	2	13	19
Tamantirto	Kasihan	12,188	5	4	9	18
Terong	Dlingo	4,625	9	4	5	18
Selopamioro	Imogiri	11,788	2	4	11	17
Sendangsari	Pajangan	9,306	8	3	6	17
Srigading	Sanden	9,104	5	5	7	17
Canden	Jetis	9,406	4	4	18	16
Mulyodadi	Bambanglipuro	11,045	7	2	7	16
Sriharjo	Imogiri	8,758	4	4	8	16
Caturharjo	Pandak	10,465	4	3	7	14
Temuwuh	Dlingo	5,776	7	3	4	14
Bangunjiwo	Kasihan	16,116	5	4	4	13
Dlingo	Dlingo	5,626	7	1	5	13
Potorono	Banguntapan	7,287	7	1	5	13
Seloharjo	Pundong	9,361	4	0	9	13
Kebonagung	Imogiri	3,204	3	2	7	12
Triwidadi	Pajangan	9,062	5	3	4	12
Gadingharjo	Sanden	3,392	2	2	7	11
Mangunan	Dlingo	4,012	7	2	2	11
Segoroyoso	Pleret	5,262	4	2	5	11
Triharjo	Pandak	10,486	4	1	5	10
Jambidan	Banguntapan	6,926	4	1	4	9
Karangtengah	Imogiri	4,228	3	0	6	9
Wirokerten	Banguntapan	7,627	3	3	3	9
Wonolelo	Pleret	3,674	2	2	5	9
Jatimulyo	Dlingo	6,459	4	3	0	7
Muntuk	Dlingo	6,741	5	1	1	7
Parangtritis	Kretek	6,333	7	0	0	7
Girirejo	Imogiri	4,058	1	1	4	6
Guwosari	Pajangan	8,020	2	2	1	5
Singosaren	Banguntapan	2,209	2	1	2	5
Tamanan	Banguntapan	6,609	5	0	0	5
Bawuran	Pleret	4,837	1	2	1	4

Source: Field work data, 1990

* Names of subdistrict capitals are printed in Italics

Annex 9.2: Incoming interaction patterns for all settlements, absolute figures and percentages, Bantul District, 1990

Settlement*	Subdistrict	incoming interactions absolute	Incoming interactions (percentages)			
			Internal	from own sub-district	from other subdistrict	Total
Regional town						
Bantul	Bantul	3,725	11	21	68	100
District towns						
Imogiri	Imogiri	1,851	12	58	30	100
Pleret	Pleret	906	26	60	14	100
Srimulyo	Piyungan	560	15	47	38	100
Sidomulyo	Bambanglipuro	516	20	2	79	100
Wijirejo	Pandak	508	39	30	31	100
Locality towns						
Banguntapan	Banguntapan	344	34	53	14	100
Trimurti	Srandakan	341	45	26	29	100
Panggungharjo	Sewon	307	62	18	20	100
Palbapang	Bantul	283	55	20	25	100
Tirtomulyo	Kretek	281	45	43	12	100
Srigading	Sanden	279	61	25	14	100
Patalan	Jetis	276	60	27	13	100
Pendowoharjo	Sewon	264	60	16	24	100
Wonokromo	Pleret	226	45	35	21	100
Caturharjo	Pandak	191	80	8	12	100
Mulyodadi	Bambanglipuro	149	66	15	19	100
Trirengo	Bantul	124	63	7	30	100
Srimartani	Piyungan	90	49	7	44	100
Rural villages `High'						
Ringinharjo	Bantul	97	89	2	9	100
Mangunan	Dlingo	79	90	1	9	100
Bangunharjo	Sewon	202	81	10	8	100
Panjangrejo	Pundong	124	66	26	8	100
Sabdodadi	Bantul	149	78	15	7	100
Canden	Jetis	139	92	1	6	100
Tamantirto	Kasihan	84	85	10	6	100
Sendangsari	Pajangan	229	62	33	6	100
Gilangharjo	Pandak	218	65	30	6	100
Sumbermulyo	Bambanglipuro	402	44	51	5	100
Donotirto	Kretek	862	30	65	5	100
Gadingharjo	Sanden	65	88	8	5	100
Murtigading	Sanden	233	46	49	4	100
Timbulharjo	Sewon	302	74	22	4	100
Sitimulyo	Piyungan	122	89	8	3	100
Seloharjo	Pundong	93	97	0	3	100
Poncosari	Srandakan	191	66	30	3	100
Singosaren	Banguntapan	98	94	3	3	100
Parangtritis	Kretek	109	96	1	3	100
Triharjo	Pandak	79	92	5	3	100

Annex 9.2 (continued)

Settlement*	Subdistrict	incoming interactions absolute	Incoming interactions (percentages)			
			Internal	from own subdistrict	from other subdistrict	Total
Kebonagung	Imogiri	123	89	9	2	100
Karangtengah	Imogiri	84	86	12	2	100
Tirtosari	Kretek	88	93	15	2	100
Srihardono	Pundong	638	44	54	2	100
Tirtonirmolo	Kasihan	278	65	33	2	100
Gadingsari	Sanden	393	46	52	2	100
Baturetno	Banguntapan	209	53	45	2	100
Triwidadi	Pajangan	109	97	1	2	100
Sriharjo	Imogiri	129	81	17	2	100
Guwosari	Pajangan	132	95	4	2	100
Jagalan	Banguntapan	67	96	3	1	100
Wirokerten	Banguntapan	74	91	8	1	100
Argosari	Sedayu	165	70	28	1	100
Tamanan	Dlingo	90	86	13	1	100
Sumberagung	Jetis	493	37	62	1	100
Bangunjiwo	Kasihan	127	90	9	1	100
Segoroyoso	Pleret	161	66	33	1	100
Rural villages 'Low'						
Karangtalun	Imogiri	401	27	73	0	100
Dlingo	Dlingo	422	39	61	0	100
Argorejo	Sedayu	411	45	55	0	100
Temuwuh	Dlingo	211	46	54	0	100
Terong	Dlingo	463	53	47	0	100
Argomulyo	Sedayu	299	62	37	0	100
Trimulyo	Jetis	276	66	34	0	100
Wonolelo	Pleret	98	67	33	0	100
Ngestiharjo	Kasihan	131	79	21	0	100
Bawuran	Pleret	51	88	12	0	100
Muntuk	Dlingo	77	90	10	0	100
Jatimulyo	Dlingo	40	90	10	0	100
Girirejo	Imogiri	42	90	10	0	100
Selopamioro	Imogiri	82	95	5	0	100
Wukirsari	Imogiri	69	96	4	0	100
Jambidan	Banguntapan	105	97	3	0	100
Potorono	Banguntapan	87	98	2	0	100
Tirtoharjo	Kretek	85	99	1	0	100
Argodadi	Sedayu	105	99	1	0	100

Source: Field work data, 1991

* Names of subdistrict capitals are printed in Italics

List of Figures

Figure 3.1:   The Region of Mopti in Mali, p. 36
Figure 3.2:   The Mopti Region: infrastructure, towns and rural production, p. 38
Figure 3.3:   Theoretical and actual flows of products from and to the Mopti Region, p. 41
Figure 3.4:   Economic activities (main and additional activities) of Mopti's working population according to sex, economic sector and residential area (in %), p. 51
Figure 3.5:   Location of main types of quarters in Mopti, p. 57

Figure 4.1:   Central Mali in its national context, p. 66
Figure 4.2:   Mali, administrative centres and districts in the research area, p. 68
Figure 4.3:   A spatial hierarchy of service centres in Central Mali, p. 74
Figure 4.4:   Potential and actual number of pupils per class, p. 81
Figure 4.5:   Distance and schooling related, results at regional level and some local examples, p. 83
Figure 4.6:   Use of medical services related to distance, a comparison of four service centres, p. 84
Figure 4.7:   The use of medical services and schools related of the educational level of the head of households, p. 87
Figure 4.8:   School use related to educational level of the head of household, according to occupational and residential aspects of the household, p. 87

Figure 5.1:   Economic and social dimensions of migration, p. 92
Figure 5.2:   Relative scores of integration levels by sector and income, p. 94
Figure 5.3:   Swaziland's major economic cores and urban settlements in 1986, p.96
Figure 5.4:   Place of birth of Swaziland born migrants living in Manzini, 1982, p. 98
Figure 5.5:   Place of birth of Nhlangano's migrants, p. 99
Figure 5.6:   Employment structures of Manzini and Nglangano, p. 101

Figure 6.1:   Huetar Norte, p. 112

Figure 7.1:   The settlement system of the State of Chihuahua, p. 130
Figure 7.2:   Migration to and from Nuevo Casas Grandes, p. 135

Figure 8.1:   The Upper Serayu Valley Region, p. 144
Figure 8.2a: A model of the urban production structure according to enterprise characteristics, p. 147
Figure 8.2b: Distribution of enterprises in Wonosobo according to number of structuring characteristics as listed in appendix, p. 147
Figure 8.3a: Matrix position of toko and warung enterprises from the sample, p. 149
Figure 8.3b: Matrix position of trade enterpises from the sample, p. 149

Figure 8.4: Model of the regional production structure and its intersectoral relations, p. 153
Figure 8.5: Two possible ways of tobacco marketing in the hinterland of Wonosobo, p. 154
Figure 8.6: Commodity flows according to the hierarchical pattern and the by-pass pattern, p. 155
Figure 8.7: A sectoral model of inter-urban relations at the national level, p. 157
Figure 8.8: Registered contractors in Kabupaten Wonosobo, 1974-1984, p. 159
Figure 8.9: The labour force according to earning capacity based on the Sayogyo norm, p. 168

Figure 9.1: Typology of subdistricts in Bantul District, p. 181
Figure 9.2: Settlement size and population density, p. 183
Figure 9.3: A hierarchy of settlements on the basis of all services, p. 184
Figure 9.4: Interactions with the City of Yogyakarta as destination, p. 186
Figure 9.5: Interactions with 'low level' rural villages as destination, p. 188
Figure 9.6: Interactions with 'high level' rural villages as destination, p. 189
Figure 9.7: Interactions with locality towns as destination, p. 191
Figure 9.8: Interactions with district towns as destination, p. 191
Figure 9.9: Interactions with regional town Bantul as destination, p. 192
Figure 9.10: Location of education and health services as actually used by households (percentages), p. 195
Figure 9.11: Use of education and health services in Bantul District (percentages), p. 197

List of Tables

Table 2.1: Enterprise matrix scoring on form and structuring characteristics, p. 28
Table 2.2: The adjusted model of the urban labourmarket, p. 30

Table 4.1: The settlement structure of Central Mali, p. 69
Table 4.2: Overview of main characteristics of the ODRs active in Central Mali, p. 71
Table 4.3: A hierarchy of service centres in Central Mali, p. 73
Table 4.4: The "extension agent density" per district and sub-district, p. 78
Table 4.5: A comparison of expected and actual numbers of patients seeking treatment at polyclinics in the research area, p. 79
Table 4.6: Average number of pupils per class in primary schools in the research area, p. 81
Table 4.7: Service use per occupation cluster (in % households per cluster), p. 86

Table 5.1: Employment by type of economic activity and by Sex, Manzini and Nhlangano (in percent of the employed of 15 years and older), p. 100
Table 5.2: Proportion of the economically active population employed amongst various categories in Manzini and Nhlangano, p. 102
Table 5.3: Integration level of migrants in Manzini and Nhlangano (in % of economically active persons over 14 years), p. 103
Table 5.4: Composition of Migrant Households in Manzini and Nhlangano, p. 105
Table 5.5: Housing situation of migrants in Nhlangano and Manzini, by period of migration (in % of heads of households), p. 106
Table 5.6: Migrants connections with the home area, Manzini (A) and Nhlangano (B), by period of arrival, p. 107

Table 6.1: Employment sectors in the labour market of Cd. Quesada, 1990, p. 118
Table 6.2: Working population of fifteen years and over in Cd. Quesada by employment sector (%), 1990, p. 119
Table 6.3: Activity status of locals and migrants in Cd. Quesada (%), 1990, p. 121
Table 6.4: Working locals and migrants by employment sector in Cd. Quesada (%), 1990, p. 122
Table 6.5: Activity status of sub-groups of migrants according to their geographical origin in Cd. Quesada (%), 1990, p. 123
Table 6.6: Sub-group of migrants by employment sector in Cd. Quesada (%), 1990, p. 124
Table 6.7: Sub-groups of working migrants in Cd. Quesada by sex (%), 1990, p. 125

Table 7.1: The changing labor force of the Municipio of Nuevo Casas Grandes, Estado de Chihuahua, 1970-1990, p. 133

Table 8.1: Population growth of the main urban centres in the Serayu Valley Region,

	1930-1980, p. 145
Table 8.2:	Estimated percentage distribution of origin of inputs according to type of enterprise in Wonosobo, p. 150
Table 8.3:	Successful enterprises in Purworejo-Klampok per sector and type of activity, p. 152
Table 8.4:	Percentage of traders at main markets in Wonosobo and Banjarnegara according to origin of main commodity sold, p. 157
Table 8.5:	Origin of customers at selected shops in Wonosobo, p. 160
Table 8.6:	Labour force participation rates, p. 161
Table 8.7:	Employment structure of Banjarnegara, Purbalingga and Wonosobo, p. 162
Table 8.8:	Percentage distribution of the labour force of the three towns according to sector and integration level in percentages, p. 164
Table 8.9:	Mean individual incomes (x Rp 1,000) per month per cell of the model in Banjarnegara, Purbalingga and Wonosobo, p. 165
Table 8.10:	Prevailing educational attainment at the various segments of the model, p. 166
Table 8.11:	Percentage shares of the economically active who never changed position within the model, per segment, p. 171
Table 8.12:	Labour mobility within the model compared, p. 171
Table 9.1:	Settlements with the highest service level per subdistrict Bantul district, 1990, p. 185
Table 9.2:	Interactions from Bantul district according to main destination Bantul district, 1991, absolute figures and percentages, p. 186
Table 9.3:	Incoming interaction patterns for the five types of settlements. Absolute figures and percentages, Bantul district, 1991, p. 187
Table 9.4:	Settlement with the highest centrality level per subdistrict, Bantul District, 1991, p. 187
Table 9.5:	Settlements in Bantul district by hierarchy and centrality, p. 193
Table 9.6:	Use of services by agro-physical zone (percentages), p. 198
Table 9.7:	Use of services by educational level of the head of the household (percentages), p. 199
Table 9.8:	Use of services by status of employment of the head of household (percentages), p. 200
Table 9.9:	Use of services by household income (percentages), p. 201
Table 10.1:	Scoring table on the strength of hinterland functions of small towns, p. 221

List of Authors

Béneker, T. (Dr.)	-	Assistant lecturer Human Geography, Latin America
Broekhuis, E.J.A. (Dr.)	-	Lecturer in Regional Development Studies, Sub-Sahara Africa
Hinderink, J. (Prof. dr.)	-	Professor (emeritus) Human Geography of Developing Countries, Sub-Sahara Africa
Huisman, H. (Dr.)	-	Senior Lecturer Rural and Regional Development Studies, South Africa and Southeast Asia
Jong, A.A. de (Dr.)	-	Lecturer Rural Development Studies, Sub-Sahara Africa
Post, C.M. van der (Dr.)	-	Senior Lecturer Regional Development Studies University of Botswana, Gaborone
Romein, A. (Dr.)	-	Staff DHV-Consultants, Amersfoort
Stoffers, J.W. (Drs.)	-	Rural Development Advisor to the Government of Botswana, Gaborone
Teeffelen, P.B.M. van (Dr.)	-	Lecturer Geographical Information Systems, Sub-Sahara Africa
Titus, M.J. (Dr.)	-	Senior Lecturer Regional Development Studies, Southeast Asia
Verkoren, O. (Dr.)	-	Assistant Professor Urban and Economic Geography, Latin America
Wouden, A.A. van der (Dr.)	-	Research fellow Regional Development Studies, Southeast Asia.